precalculus algebra and trigonometry

precalculus algebra and trigonometry

DANIEL D. BENICE

Montgomery College
Rockville, Maryland

PRENTICE-HALL, INC., Englewood Cliffs, New Jersey

Library of Congress Cataloging in Publication Data

BENICE, DANIEL D
Precalculus algebra and trigonometry.

Includes index.
1. Algebra. 2. Trigonometry. I. Title.
QA152.2.B46   512′.1   75-22236
ISBN 0-13-695072-8

10 9 8 7 6 5 4 3 2 1

Printed in the United States of America

PRENTICE-HALL INTERNATIONAL, INC., *London*
PRENTICE-HALL OF AUSTRALIA, PTY. LTD., *Sydney*
PRENTICE-HALL OF CANADA, LTD., *Toronto*
PRENTICE-HALL OF INDIA PRIVATE LIMITED, *New Delhi*
PRENTICE-HALL OF JAPAN, INC., *Tokyo*
PRENTICE-HALL OF SOUTHEAST ASIA (PTE.) LTD., *Singapore*

The Spirit of '76

contents

preface

This book is primarily intended to serve students of algebra and trigonometry who plan to continue on to calculus. In addition to the
standard topics in algebra, trigonometry, and elementary functions,
many topics, examples, and exercises have been designed to compare
with the kinds of algebra problems students face in calculus courses.
A brief look at the first two chapters and the last chapter, for example,
will confirm this. On closer examination you'll find a variety of truly
calculus-oriented examples and exercises, including:

(a) Complex fractions of the type arising from using the difference
   quotient.

(b) Factoring involving negative and fractional exponents. These
   are the same types that often arise after taking the derivative
   of a product. Later, the process is extended to obtain a fraction in order to see where the expression is zero or undefined.
   Such a procedure is needed for maximum/minimum calculus
   problems.

(c) Fractions in which the *numerator* must be rationalized. In calculus, this occurs when applying the difference quotient definition of derivative to $f(x) = \sqrt{x}$.

(d) Inequalities involving products, quotients, and polynomials, and their application to determining the implied domain and range of relations and functions of the type used in calculus. Many must be solved explicitly for x or y first.

(e) Inequalities of the type students use with the epsilon-delta definition of *limit*.

(f) Calculations using two different types of difference quotients.

(g) Geometric problems like those that students encounter in maximum/minimum and related rates settings.

(h) A variety of exercises involving e.

(i) Polar coordinates.

(j) Partial fractions.

(k) Iteration.

(l) A short chapter on calculus, which includes intuitive introductions to secant lines, tangent lines, limits, derivatives, maximum/minimum, and the meaning and use of such terms as *theorem*, *corollary*, and *if* and *only if*.

The text is student-oriented. Examples and explanations are intended to be clear to the average student. Some theory has been omitted, but definitions, theorems, and proofs appear as needed. Answers are given to enough problems to enable the reader to find and correct his errors.

For the student, Richard Detmer has prepared a useful study guide. It contains review of basic concepts, supplementary ideas, solutions to selected text exercises, and self-tests for each chapter. Self-test answers are given with a key specifying where in the text that material is covered.

For the instructor, a manual containing outlines and answers is available.

The writing and rewriting of this book has been an enjoyable educational experience. I would like to acknowledge the invaluable assistance of several mathematicians who are directly responsible for many improvements made since the original first draft. They are:

Donald R. Burleson (Middlesex Community College), R. A. Close (Pan American University), Richard C. Detmer (University of Tennessee at Chattanooga), Edward Doran (Community College of Denver, North Campus), Samuel Goldberg (Oberlin College), Joseph J. Hansen (Northeastern University), and Kenneth Hoffman (Massachusetts Institute of Technology).

DANIEL D. BENICE

precalculus algebra
and
trigonometry

chapter one

real and complex numbers

1.1 THE REAL NUMBER SYSTEM

This first chapter provides you with a brief review of some basic algebra. It introduces definitions, concepts, and skills that will be needed later in the book and in calculus. Inequalities are emphasized in Sections 1.6 and 1.7 because of their extensive application later in the book and throughout calculus. Once you have learned the material presented in Chapter 1, you should be prepared to begin the study of functions, which starts in Chapter 2.

One simple number system consists of the *positive integers*: 1, 2, 3, 4, etc. and their arithmetic. The numbers in this system are also called the *counting numbers* or *natural numbers*. Listed below are some of the named properties that positive integers possess with respect to the operations of addition and multiplication. The letters a, b, c, x, and y are used here to represent any positive integers.

Closure Properties

If any two positive integers are added, their sum is a unique positive integer. If any two positive integers are multiplied, their product is a unique positive integer.

Associative Properties

Addition: $(a + b) + c = a + (b + c)$

Multiplication: $(ab)c = a(bc)$

Addition and multiplication are binary operations in that they are applied to two numbers to yield a unique number. Three numbers cannot be added or multiplied at the same time. First two of them must be added (or multiplied); then the third can be added to (or multiplied by) that result. Therefore, when the sum or product of three numbers is indicated, parentheses should be used to specify the way in which the numbers are to be combined. Thus, $3 + 4 + 5$ can be written as $(3 + 4) + 5$, which becomes $7 + 5$ and then 12. The sum $3 + 4 + 5$ can also be considered as $3 + (4 + 5)$, which becomes $3 + 9$ and then 12. The sums are equal, but in the first case the 4 is *associated* with the 3 and added to it, whereas in the second case the 4 is associated with the 5 and added to it.

Commutative Properties

Addition: $a + b = b + a$

Multiplication: $a \cdot b = b \cdot a$

The sum $2 + 5$ is the same as the sum $5 + 2$. The product $4 \cdot 3$ is the same as the product $3 \cdot 4$. A product such as $y \cdot 7$ can be written as $7 \cdot y$, if desired.

Distributive Properties

$$a(b + c) = a \cdot b + a \cdot c$$
$$(b + c)a = b \cdot a + c \cdot a$$

The product $8(4 + 3)$ can be written as $8 \cdot 4 + 8 \cdot 3$. Similarly,

$$5(x + y) = 5x + 5y$$
$$(6 + 9)x = 6x + 9x$$

The distributive properties are used in combining like terms, factoring, and expanding products. Several examples follow.

Example 1. *Combine like terms:* $5x + 9x$.

$$5x + 9x = (5 + 9)x = 14x \quad \checkmark$$

Example 2. *Factor* $4(a + b) + 7(a + b)$.

$$4(a + b) + 7(a + b) = (4 + 7)(a + b) = 11(a + b) \quad \checkmark$$

Example 3. *Factor* $x(a + b) + y(a + b)$.

$$x(a + b) + y(a + b) = (x + y)(a + b) \quad \checkmark$$

Subtraction can be defined in terms of addition. The *difference* $a - b$ is the number d such that $a = b + d$. The system of positive integers is not *closed* under subtraction; that is, subtraction does not satisfy the closure property.

The number *zero* is introduced in order to define the difference $a - a$. Zero is the number that must be added to a to produce a. The positive integers and the number zero together form the *whole numbers*.

Still it is possible to have a difference $a - b$ which is neither a positive integer nor zero. *Negative integers* are introduced to handle subtractions such as $7 - 10$. The negative integers are, in descending order, $-1, -2, -3, -4$, etc.

The *integers* consist of positive integers, zero, and the negative integers. The integers are closed under addition, multiplication, and subtraction.

The number zero is sometimes called the *additive identity*. If zero is added to any number x, the sum is x. In symbols, $x + 0 = x$.

The solution to the equation $2 + x = 0$ is -2. The number -2 is called the *negative* or *opposite* or *additive inverse* of 2. Similarly, 2 is the additive inverse of -2. In general, if $a + x = 0$, then $x = -a$, the additive inverse of a.

Example 4. *Factor* $x^2 - y^2$.

$$x^2 - y^2 = x^2 - xy + xy - y^2 \qquad \begin{cases} \text{after adding 0 in the form} \\ -xy + xy \end{cases}$$

$$= x(x - y) + y(x - y) \qquad \begin{cases} \text{after factoring } x \text{ from the} \\ \text{first two terms and } y \text{ from} \\ \text{the other two terms} \end{cases}$$

$$= (x + y)(x - y) \quad \checkmark$$

This, of course, is the standard factoring for the *difference of two squares*. The form should be memorized rather than reworked each time.

$$\boxed{x^2 - y^2 = (x + y)(x - y)}$$

As another example,

$$x^2 - 25 = (x + 5)(x - 5)$$

The *sum* or *difference of two cubes* can also be factored.

$$x^3 - y^3 = (x - y)(x^2 + xy + y^2)$$
$$x^3 + y^3 = (x + y)(x^2 - xy + y^2)$$

Example 5. *Expand* $(x + 2)(x + 3)$.

$$
\begin{aligned}
(x + 2)(x + 3) &= (x + 2)x + (x + 2)3 \\
&= x^2 + 2x + 3x + 6 \\
&= x^2 + 5x + 6 \quad \checkmark
\end{aligned}
$$

You probably know a shortcut for performing such multiplication. The example illustrates the distributive property.

Example 6. *Factor* $x^3 - 16x$.

$$x^3 - 16x = x(x^2 - 16) = x(x + 4)(x - 4) \quad \checkmark$$

Division can be defined in terms of multiplication. The *quotient* a/b is the number q such that $a = bq$. It is possible to have a quotient that is not an integer. For example, $7/3$ is not an integer; $7 = 3q$ does not have an integer solution. In order to permit division of integers, the rational numbers are defined. A *rational number* is any number that can be written in the form a/b, where a and b are integers and b is not zero.

Let's examine the reason that b cannot be zero in the fraction a/b. By definition $a/b = q$ if $a = bq$. If $b = 0$, then $a = bq$ becomes $a = 0$. Thus, if $b = 0$, then a must be zero also. In that case q cannot be determined, since $a = bq$ becomes $0 = 0 \cdot q$. Any value for q will satisfy the equation. The quotient $0/0$ is appropriately called *indeterminate*; the quotient cannot be determined. The quotient $a/0$, $a \neq 0$, is *undefined*, because there is no value for a that yields a quotient.

Here are some examples of rational numbers and an a/b form which shows that they are indeed rational.

$$7 \quad \longrightarrow \frac{7}{1}$$

$$2.3 \quad \longrightarrow \frac{23}{10}$$

$$1\frac{7}{8} \longrightarrow \frac{15}{8}$$

$$-3 \longrightarrow \frac{-6}{2}$$

$$0.33\bar{3} \longrightarrow \frac{1}{3}$$

$$0 \longrightarrow \frac{0}{5}$$

The bar over the 3 indicates that the 3 repeats indefinitely. Any repeating decimal is a rational number. (See problem 11. Exercise Set A.)

Not all numbers are rational. For example, $\sqrt{2}$ is not rational; it is an *irrational number*. It can be proved that $\sqrt{2}$ is not rational by showing that $\sqrt{2}$ cannot be written in the form a/b, where both a and b are integers. The proof is by contradiction. Suppose there are integers a' and b' such that $a'/b' = \sqrt{2}$. Let a/b represent the completely reduced form of a'/b'. Then $a = \sqrt{2}\,b$ or $a^2 = 2b^2$. Since $a^2 = 2b^2$, a^2 is a multiple of 2. This means that $a \cdot a$ is a multiple of 2, which in turn means that a must be a multiple of 2. Thus, $a = 2c$ for some integer c. Substituting $2c$ for a in $a^2 = 2b^2$ yields $2c \cdot 2c = 2b^2$, which shows that b^2 is a multiple of 2. Thus, b is a multiple of 2. In other words, both a and b are multiples of 2, so that a/b is not completely reduced, as assumed. This is contrary to the original statement and thus shows that $\sqrt{2}$ cannot be written as a quotient of integers, that $\sqrt{2}$ is not rational.

Similarly, the square root of any positive number that is not a perfect square is an irrational number; for example, $\sqrt{3}$, $\sqrt{12}$, and $\sqrt{20}$ are irrational. Another common irrational number is π. The value of π is often *approximated* as $\pi \doteq 3.14$, but this number is not exactly π; it is a rational approximation of this irrational number. Note that the symbol $\doteq$ is used to mean approximately equal to. When an irrational number appears in an expression that must be evaluated, such rational approximations are useful even though they are not the exact values.

Addition, multiplication, division, and reduction of rational numbers are defined next. The numbers a, b, c, and d are integers.

1. $\dfrac{a}{d} + \dfrac{b}{d} = \dfrac{a+b}{d}$ $\qquad\qquad (d \neq 0)$

2. $\dfrac{a}{b} \cdot \dfrac{c}{d} = \dfrac{ac}{bd}$ $\qquad\qquad (b, d \neq 0)$

3. $\dfrac{a}{b} + \dfrac{c}{d} = \dfrac{a}{b} \cdot \dfrac{d}{d} + \dfrac{c}{d} \cdot \dfrac{b}{b}$

$$= \dfrac{ad}{bd} + \dfrac{bc}{bd} = \dfrac{ad + bc}{bd} \qquad (b, d \neq 0)$$

4. $\dfrac{ac}{bc} = \dfrac{a}{b}$ \qquad\qquad\qquad\qquad $(b, c \neq 0)$

5. $\dfrac{a}{b} \div \dfrac{c}{d} = \dfrac{ad}{bc}$ \qquad\qquad\qquad $(b, c, d \neq 0)$

Example 7. *Combine* $\dfrac{3x}{5ab} + \dfrac{7}{4c}$.

The common denominator is $5ab \cdot 4c$ or $20abc$. Thus,

$$\dfrac{3x}{5ab} + \dfrac{7}{4c} = \dfrac{3x}{5ab} \cdot \dfrac{4c}{4c} + \dfrac{7}{4c} \cdot \dfrac{5ab}{5ab}$$

$$= \dfrac{12cx}{20abc} + \dfrac{35ab}{20abc}$$

$$= \dfrac{12cx + 35ab}{20abc} \quad \checkmark$$

Example 8. *Reduce* $\dfrac{x^2 + 7x + 12}{x^2 - x - 20}$.

The numerator and denominator can be factored and the common factor divided out according to property 4.

$$\dfrac{x^2 + 7x + 12}{x^2 - x - 20} = \dfrac{(x + 3)(x + 4)}{(x - 5)(x + 4)} = \dfrac{x + 3}{x - 5} \quad \checkmark$$

Example 9. *Multiply* $\dfrac{x + 4}{5} \cdot \dfrac{2x + 1}{y}$.

$$\dfrac{x + 4}{5} \cdot \dfrac{2x + 1}{y} = \dfrac{(x + 4)(2x + 1)}{5y} = \dfrac{2x^2 + 9x + 4}{5y} \quad \checkmark$$

Often, a factored form is more useful than a multiplied out form. The result here can be left as the fraction

$$\dfrac{(x + 4)(2x + 1)}{5y} \quad \checkmark$$

Example 10. *Divide* $\dfrac{x^2 - 9}{x} \div \dfrac{2x + 6}{8}$.

$$\frac{x^2 - 9}{x} \div \frac{2x + 6}{8} = \frac{x^2 - 9}{x} \cdot \frac{8}{2x + 6}$$

$$= \frac{(x + 3)(x - 3)}{x} \cdot \frac{2 \cdot 4}{2(x + 3)} = \frac{(x - 3)4}{x} \quad \checkmark$$

Example 11. *Simplify* $\dfrac{\dfrac{x}{y} + 5}{2 + \dfrac{1}{y}}$.

The terms x/y and 5 can be combined as

$$\frac{x}{y} + 5 = \frac{x}{y} + \frac{5}{1} = \frac{x}{y} + \frac{5y}{y} = \frac{x + 5y}{y}$$

Similarly,

$$2 + \frac{1}{y} = \frac{2y + 1}{y}$$

Thus,

$$\frac{\dfrac{x}{y} + 5}{2 + \dfrac{1}{y}} = \frac{\dfrac{x + 5y}{y}}{\dfrac{2y + 1}{y}} = \frac{x + 5y}{y} \div \frac{2y + 1}{y}$$

$$= \frac{x + 5y}{y} \cdot \frac{y}{2y + 1}$$

$$= \frac{x + 5y}{2y + 1} \quad \checkmark$$

Example 12. *Combine* $\dfrac{5}{2x - 8} + \dfrac{7}{x^2 - 2x - 8}$.

$$\frac{5}{2x - 8} + \frac{7}{x^2 - 2x - 8} = \frac{5}{2(x - 4)} + \frac{7}{(x - 4)(x + 2)}$$

The common denominator is $2(x - 4)(x + 2)$.

$$= \frac{5}{2(x - 4)} \cdot \frac{(x + 2)}{(x + 2)} + \frac{7}{(x - 4)(x + 2)} \cdot \frac{2}{2}$$

$$= \frac{5(x + 2) + 7 \cdot 2}{2(x - 4)(x + 2)}$$

$$= \frac{5x + 24}{2x^2 - 4x - 16} \quad \checkmark$$

One final note on fractions. When an equation contains fractions, it is usually best to begin solving it by eliminating all the fractions. This can be done by multiplying both sides of the equation (all terms) by the common denominator of all fractions involved. The resulting equation will contain no fractions and should be much easier to solve.

Example 13. *Solve* $\dfrac{5}{x+1} - \dfrac{3}{20} = \dfrac{9}{5}$.

The common denominator is $20(x+1)$. If both sides of the equation are multiplied by $20(x+1)$, we get

$$20(x+1) \cdot \frac{5}{x+1} - 20(x+1) \cdot \frac{3}{20} = 20(x+1) \cdot \frac{9}{5}$$

or

$$100 - 3(x+1) = 36(x+1)$$

or

$$100 - 3x - 3 = 36x + 36$$

Then

$$61 = 39x$$

Finally,

$$x = \frac{61}{39} \quad \checkmark$$

1.2 THE REAL NUMBER PLANE

The *real numbers* consist of the rational numbers together with the irrational numbers. In fact, the real numbers consist of all numbers that are either positive, zero, or negative. The *real number line* is used to illustrate and compare real numbers.

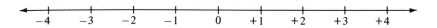

Positive numbers are located to the right of zero; negative numbers are located to the left of zero. Integers are located by moving out equal distances from zero. To the right, one unit is $+1$, two units $+2$, three units $+3$, etc. To the left, one unit is -1, two units -2, etc. Any rational number that is not an integer can be written as a/b ($b \neq 0$) and located between two integers. For example, the number $1.3 = \frac{13}{10}$ can be located on the number line by dividing the interval between 1 and 2 into 10 equal parts and counting over 3 of those parts.

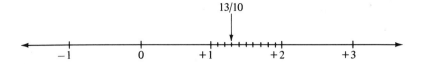

Irrational numbers can also be located on the real number line. To locate $\sqrt{2}$, construct a square having at its base zero at one vertex and 1 at the other vertex. Use of the Pythagorean theorem ($c^2 = a^2 + b^2$, c = hypotenuse) yields $\sqrt{2}$ as the length of the diagonal.

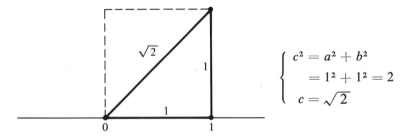

$$\begin{cases} c^2 = a^2 + b^2 \\ = 1^2 + 1^2 = 2 \\ c = \sqrt{2} \end{cases}$$

With center at 0, draw an arc with radius equal to the diagonal of the square. It crosses the number line at $\sqrt{2}$, since the length of the diagonal is $\sqrt{2}$.

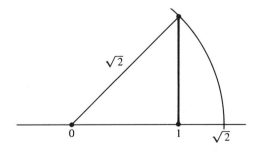

If two number lines are positioned perpendicular to each other and joined at their zeros, the result is the coordinate plane. Traditionally, the horizontal line is called the *x axis* and the vertical line is the *y axis*. Other letters are sometimes used.

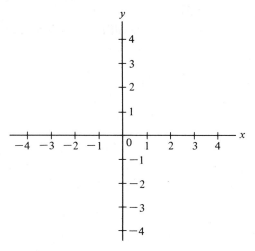

Points are specified by ordered pairs (x, y) of real numbers. The first coordinate, x, specifies a right (positive) or left (negative) direction. The second coordinate, y, specifies an up (positive) or down (negative) direction. All distances are measured from the origin $(0, 0)$. Next, several points are shown plotted. The quadrants are numbered I, II, III, and IV counterclockwise beginning in the top right quadrant.

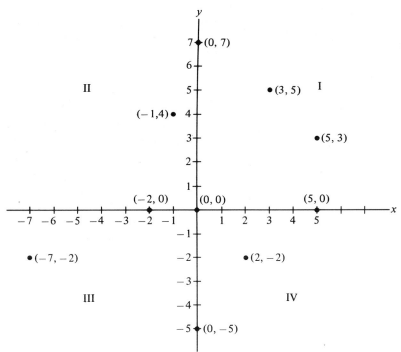

Note that $(3, 5) \neq (5, 3)$. They are different points. Points are *ordered* pairs. The order in which the coordinates are written is critical.

The real numbers, together with the operations of addition and/or multiplication, satisfy several important properties. The closure, associative, commutative, and distributive properties studied earlier apply to all real numbers, not just positive integers. There are still other basic properties satisfied by real numbers. Some of these are explained in the exercises. The properties of signed numbers which you learned in elementary algebra apply to all real numbers. The product or quotient of two positive numbers or two negative numbers is a positive number. The product or quotient of two unlike signed numbers is a negative number.

EXERCISE SET A

1. For each real number listed, use as many of the following descriptions that apply: rational, irrational, integer, whole, natural.

   ⋆(a) 6   (b) -9 ⋆(c) $\dfrac{2}{5}$

   (d) $\dfrac{7}{-3}$ ⋆(e) $\sqrt{3}$ (f) 1.962

   ⋆(g) 3.14   (h) $-\sqrt{2}$ ⋆(i) $\sqrt{1}$

   (j) 0   (k) $\sqrt{\dfrac{9}{4}}$ ⋆(l) $3\tfrac{5}{16}$

2. The additive identity is 0, since if 0 is added to any number x, the result is x; x retains its identity. What is the *multiplicative identity*, and why?

3. The additive inverse of any number x is $-x$, since when the number and its inverse are added, they produce the identity. What is the *multiplicative inverse* (or *reciprocal*) of the nonzero rational x/y, and why? Also, does it matter whether or not x/y is zero; that is, does 0 have a multiplicative inverse?

4. Factor each expression completely.

   ⋆(a) $2x + ax$ ⋆(b) $y^3 - 7y^2$

   (c) $3x^2 + 6x$ ⋆(d) $x^2 - 100$

   (e) $m^2 - n^2$ ⋆(f) $2(x + y) + m(x + y)$

   (g) $w\sqrt{5} + x\sqrt{5} + y\sqrt{5}$ ⋆(h) $a(a - b) - b(a - b)$

   ⋆(i) $x^2 + 7x + 6$ (j) $m^2 - 9m + 20$

   (k) $y^2 + 5y - 14$ ⋆(l) $2b^2 - 8$

⋆Answers to starred problems are to be found at the back of the book.

(m) $2x^2 + 2x - 4$

★(o) $a^5 - 9a^3$

★(q) $x^4 - 1$

★(s) $m^3 - 27$

(u) $m^2n - mn^2 - mn$

(n) $n^3 - 4n^2 - 12n$

(p) $x(x^2 - 1) + 3(x^2 - 1)$

(r) $a^3 - x^3$

(t) $8n^3 + 1$

(v) $4a^2x^2 - 4ax + 1$

5. Perform the specified operation and simplify.

★(a) $\dfrac{x}{a} + \dfrac{y}{b}$

(b) $\dfrac{x}{a} + \dfrac{a}{x}$

★(c) $\dfrac{5}{xy} + \dfrac{7x}{3y}$

★(d) $\dfrac{4}{x+2} - \dfrac{3}{x-5}$

(e) $\dfrac{3x}{x^2 - 25} + \dfrac{x+1}{x^2 + 5x}$

★(f) $\dfrac{2}{x^2 - 1} + \dfrac{5}{x^2 - 7x + 10}$

★(g) $\dfrac{x}{y} \cdot \dfrac{2y}{x+1}$

(h) $\dfrac{x+3}{9} \cdot \dfrac{6x}{6x+10}$

★(i) $\dfrac{x^2 - 16}{x^2 + 6x + 8} \cdot \dfrac{x^2 + 2x}{x^3}$

(j) $\dfrac{2x+10}{4x-12} \cdot \dfrac{x^2 - 8x + 15}{x^2 - 25}$

★(k) $\dfrac{3x}{2y} \div \dfrac{x+1}{10y^2 + y}$

(l) $\dfrac{x-y}{y-x} \div \dfrac{x}{y}$

★(m) $\dfrac{x^2 + 2x}{5x} \div \dfrac{x^2 - 4}{10}$

(n) $\dfrac{x^2 - 6x + 9}{x^2 + 2x - 15} \div \dfrac{x^2 - 9}{5x}$

★(o) $\dfrac{2 + \dfrac{1}{x}}{3 - \dfrac{1}{x}}$

★(p) $\dfrac{\dfrac{x}{y} + \dfrac{y}{x}}{\dfrac{x}{y}}$

★(q) $\dfrac{2 + \dfrac{x}{y}}{\dfrac{x}{y} + 4}$

(r) $\dfrac{\dfrac{n+1}{n}}{\dfrac{2}{n} + 9}$

★(s) $\dfrac{\dfrac{2t+1}{t} + t}{\dfrac{1}{t} - t}$

(t) $\dfrac{\dfrac{m}{n} + \dfrac{n}{m}}{\dfrac{1}{n} - \dfrac{1}{m}}$

★(u) $\dfrac{\dfrac{1}{(x+h)-1} - \dfrac{1}{x-1}}{h}$

(v) $1 - \dfrac{1}{1 - \dfrac{1}{x}}$

6. Solve each equation.

★(a) $7x + 3 = 19$

(b) $5(t+4) - 2 = 0$

★(c) $3x - (4 - 3x) = 15x + 14$

(d) $\frac{1}{2}x + 16 = 4 - 3x$

★(e) $6 + \dfrac{x}{7} = 9$

★(f) $\dfrac{3x+2}{5} + \dfrac{3}{4} = 1$

(g) $\dfrac{x}{4} + \dfrac{x+1}{3} = 5$

(h) $\dfrac{x}{x+2} + \dfrac{1}{5} + \dfrac{5}{x+2} = 0$

⋆(i) $\dfrac{3}{x-1} = \dfrac{5}{x+2}$

(j) $\dfrac{m}{m-3} + \dfrac{2}{7} = \dfrac{3}{2m-6}$

⋆(k) $\dfrac{x}{x+1} + \dfrac{x}{x+5} = 2$

7. The Internal Revenue Service suggests that a common error on filed tax returns is the interchanging of two adjacent digits; for example, 2583 might be incorrectly written as 2853 or as 2538. Such an error is relatively easy to spot, because the difference between the correct number and the incorrect number is divisible by 9. In other words,

$$\text{Correct} - \text{incorrect} = 9 \cdot (\text{some integer})$$

Consider a two-digit number whose digits are m and n. Represent the number and its incorrect form. Then prove that their difference is equal to nine times some integer.

8. Prove each of the following statements by multiplying out the right side and simplifying it.
   (a) $x^3 - y^3 = (x - y)(x^2 + xy + y^2)$
   (b) $x^3 + y^3 = (x + y)(x^2 - xy + y^2)$

9. Show in steps how the fraction

$$\frac{-x^3}{x-4}$$

can be changed to the form

$$\frac{x^3}{4-x}$$

10. Can you determine what is wrong with the following proof that $1 = 2$? Let $a = b$. Then $a^2 = ab$ and $a^2 - b^2 = ab - b^2$. So

$$(a + b)(a - b) = b(a - b), \text{ or } a + b = b$$

By substitution, $b + b = b$ or $2b = b$ or $2b = 1b$. Thus, $2 = 1$.

11. To demonstrate that they are indeed rational numbers, repeating decimals can be changed to fractions by the process demonstrated here. To change $x = 0.16\overline{16}$ to a fraction, multiply it by 100 in order to align the repeating portions

$$x = 0.16\overline{16}$$
$$100x = 16.16\overline{16}$$

Now subtract the smaller number x from the larger number $100x$. Since the repeating portions are aligned, the subtraction leaves simply

$$99x = 16, \quad \text{or } x = \frac{16}{99}$$

Use this process to change each repeating decimal to a fraction.

★(a) $0.73\overline{73}$ ★(b) $0.55\overline{5}$ (c) $0.691\overline{691}$
★(d) $0.53\overline{33}$ (e) $1.212\overline{121}$ (f) $5.316161\overline{6}$

12. The system of positive integers is not closed under subtraction. Give a numerical example of two positive integers a and b such that $a - b$ is not a positive integer. Can you find positive integers a and b such that $a - b$ is a positive integer? Are the positive integers closed under division? Explain.

★13. Identify the property used.

(a) $x \cdot 3 = 3x$ (b) $9x + mx = (9 + m)x$
(c) $(4 + x) + y = 4 + (x + y)$ (d) $x + 2 = 2 + x$
(e) $5(x + 2) = 5x + 10$

14. Prove each statement.

(a) $(a + b)^2 \neq a^2 + b^2$ (b) $(a - b)^2 \neq a^2 - b^2$
(c) $(a + b)^3 \neq a^3 + b^3$ (d) $(a - b)^3 \neq a^3 - b^3$

1.3 INTEGER EXPONENTS

This section presents a brief review of the notation and properties of exponents. Letters used represent any real numbers, unless otherwise restricted.

$$4^3 \text{ is the same as } 4 \cdot 4 \cdot 4$$
$$2^5 \text{ is the same as } 2 \cdot 2 \cdot 2 \cdot 2 \cdot 2$$

In general, for any positive integer n,

$$x^n = x \cdot x \cdot x \cdots x \qquad (n \text{ factors of } x)$$

Consider the product $c^4 \cdot c^5$.

$$c^4 \cdot c^5 = (c \cdot c \cdot c \cdot c)(c \cdot c \cdot c \cdot c \cdot c) = c^{4+5} = c^9$$

In general,

$$x^m \cdot x^n = x^{m+n} \qquad (m, n \text{ positive integers})$$

The expression $(c^2)^3$ is the same as $(cc)^3$ or $(cc)(cc)(cc)$ or $c^{2 \cdot 3} = c^6$. In general,

$$(x^m)^n = x^{mn} \qquad (m, n \text{ positive integers})$$

Next, consider

$$(ab)^3 = (ab)(ab)(ab) = a^3 b^3$$

In general,

$$(x \cdot y)^m = x^m y^m \qquad (m \text{ a positive integer})$$

Similarly,

$$\left(\frac{a}{b}\right)^3 = \frac{a}{b} \cdot \frac{a}{b} \cdot \frac{a}{b} = \frac{a^3}{b^3}$$

which leads to the generalization

$$\left(\frac{x}{y}\right)^m = \frac{x^m}{y^m} \qquad (m \text{ a positive integer})$$

There are three possible cases for the fraction $\dfrac{x^m}{x^n}$. Each case is demonstrated next in an example.

$$\frac{x^5}{x^3} = \frac{x \cdot x \cdot x \cdot x \cdot x}{x \cdot x \cdot x} = x^2 \quad \text{or} \quad x^{5-3}$$

$$\frac{x^5}{x^5} = \frac{x \cdot x \cdot x \cdot x \cdot x}{x \cdot x \cdot x \cdot x \cdot x} = 1$$

$$\frac{x^3}{x^5} = \frac{x \cdot x \cdot x}{x \cdot x \cdot x \cdot x \cdot x} = \frac{1}{x^2} \quad \text{or} \quad \frac{1}{x^{5-3}}$$

The generalization:

$$\frac{x^m}{x^n} = \begin{cases} x^{m-n} & \text{if } m \text{ is larger than } n \\ 1 & \text{if } m \text{ is equal to } n \\ \dfrac{1}{x^{n-m}} & \text{if } m \text{ is smaller than } n \end{cases}$$

We define

$$x^0 = 1 \qquad \text{for } x \neq 0$$

This is consistent with $x^m \cdot x^n = x^{m+n}$, since x^0 acts as 1 in

$$x^0 \cdot x^n = x^{0+n} = x^n$$

Next we define negative exponents.

$$x^{-n} = \frac{1}{x^n} \qquad x \neq 0$$

This, too, is consistent with $x^m \cdot x^n = x^{m+n}$. Using $m = -n$ yields

$$x^{-n} \cdot x^n = x^{-n+n} = x^0 = 1$$

Letting x^{-n} equal $1/x^n$ also yields 1, as

$$\frac{1}{x^n} \cdot x^n = \frac{x^n}{x^n} = 1$$

Thus, $x^{-n} = 1/x^n$.

We shall give next several examples that show how these properties of exponents can be used to simplify algebraic expressions.

Example 14. *Simplify $(x^2 y^5)^4$.*

$$(x^2 y^5)^4 = (x^2)^4 (y^5)^4 = x^8 y^{20} \quad \checkmark$$

Example 15. *Simplify $\dfrac{(5x)^0 y^{10} z^9}{y^4 z^{12}}$.*

$$(5x)^0 = 1 \qquad \frac{y^{10}}{y^4} = y^6 \qquad \frac{z^9}{z^{12}} = \frac{1}{z^3}$$

so

$$\frac{(5x)^0 y^{10} z^9}{y^4 z^{12}} = \frac{y^6}{z^3} \quad \checkmark$$

Although z^9/z^{12} can be considered as z^{-3}, and the result of this example left as $y^6 z^{-3}$, it is preferable to avoid negative exponents in the final simplified result.

A negative power cannot be computed directly; it must be changed to a positive power in order to be evaluated.

Example 16. *Simplify* $(x^2 x^{-7})^{-3}$.

$$(x^2 x^{-7})^{-3} = (x^{-5})^{-3} = x^{15} \quad \checkmark$$

Example 17. *Simplify* $(4x^0)^2(x^{-2})^3$.

$$(4x^0)^2(x^{-2})^3 = (4 \cdot 1)^2(x^{-6}) = 16 \cdot \frac{1}{x^6} = \frac{16}{x^6} \quad \checkmark$$

Example 18. *Simplify* $(5x^3 y^2)(2x^8 y^7)$.

$$(5x^3 y^2)(2x^8 y^7) = 5 \cdot 2 \cdot x^3 \cdot x^8 \cdot y^2 \cdot y^7 = 10x^{11} y^9 \quad \checkmark$$

Example 19. *Simplify* $(x^m + y^n)(x^{-m} + y^{-n})$.

$$(x^m + y^n)(x^{-m} + y^{-n}) = x^m x^{-m} + x^m y^{-n} + y^n x^{-m} + y^n y^{-n}$$

$$= 1 + \frac{x^m}{y^n} + \frac{y^n}{x^m} + 1$$

$$= 2 + \frac{x^m}{y^n} + \frac{y^n}{x^m}$$

The common denominator for all three fractions is $x^m y^n$. They can be added as

$$= \frac{2x^m y^n}{x^m y^n} + \frac{x^m x^m}{x^m y^n} + \frac{y^n y^n}{x^m y^n}$$

$$= \frac{2x^m y^n + x^{2m} + y^{2n}}{x^m y^n} \quad \checkmark$$

Example 20. *Simplify* $\dfrac{3x^{-2} y^5 z^8}{6x^4 y^3 z^8}$.

$$\frac{3}{6} = \frac{1}{2} \qquad \frac{x^{-2}}{x^4} = \frac{1}{x^2} \cdot \frac{1}{x^4} = \frac{1}{x^6} \qquad \frac{y^5}{y^3} = \frac{y^2}{1} \qquad \frac{z^8}{z^8} = 1$$

Thus,

$$\frac{3x^{-2} y^5 z^8}{6x^4 y^3 z^8} = \frac{1}{2} \cdot \frac{1}{x^6} \cdot \frac{y^2}{1} \cdot 1 = \frac{y^2}{2x^6} \quad \checkmark$$

EXERCISE SET B

1. Simplify each expression completely. Eliminate all negative and zero exponents.

⋆(a) $6m^7m^8 \cdot 2m^2$

⋆(b) $(4x^3y^5)(x^9y^2)$

⋆(c) $(x^{-2}y^{-3})^{-5}$

(d) $(x^{-3}y^{-2})(x^3y^{-1})$

(e) $(2x^4y)(-y^3z^7)^2$

⋆(f) $(1 - x^0)^7 + (2x^0)^5 + (x^0)^4$

(g) $(x^0 + 2^3)^2$

(h) $(5x^{-3}y^2)^2$

⋆(i) $\dfrac{x^5y^{10}}{2x^{-3}y^4}$

⋆(j) $\dfrac{14n^6p^{-2}}{2q^2r^{-5}s^{-2}}$

⋆(k) $\left(\dfrac{x^3y^2}{x^5y}\right)^4$

(l) $\left(\dfrac{x^{-9}}{4x^2}\right)^3$

(m) $\left(\dfrac{3n^2}{4n^{-5}}\right)^{-2}$

⋆(n) $\dfrac{a^2}{b} + \dfrac{b^2}{a}$

⋆(o) $c^{-3}d + c^2d^{-2}$

⋆(p) $\left(\dfrac{a^0 - b^0}{a^0 + b^0}\right)^{-2}$

(q) $\dfrac{x^2}{y^{-1}} + \dfrac{y}{x^{-2}}$

(r) $7(c^0d^4e^7)^0$

⋆(s) $a^{-1} + b^{-1}$

(t) $\dfrac{x^nx^{2n}x^1}{x^4}$ $(n \geq 2)$

(u) $(a + b)(a^2 - b^2)^{-1}$

⋆(v) $(x^{-1} + y^{-1})^{-1}$

⋆(w) $(a^x + b^y)(a^{-x} - b^{-y})$

(x) $\dfrac{\frac{1}{xy}}{x^{-1} + y^{-1}}$

⋆(y) $\left(\dfrac{x^{-2}}{y} + \dfrac{y^{-2}}{x}\right)^{-1}$

⋆(z) $\dfrac{(x^2 + 1)^0}{(x^2 - 1)^2} \cdot \dfrac{1 - x^2}{x^2 + 1}$

2. Simplify each expression completely.

⋆(a) $\dfrac{x^{-1} - x^{-2}}{x}$

(b) $\dfrac{x}{x^{-1} - x^{-2}}$

⋆(c) $\dfrac{x^{-1}}{x^{-1} + x^{-2}}$

(d) $(x + y^{-2})^{-1}$

(e) $\dfrac{(x + 1)^{-1} + (x - 1)^{-1}}{x(x + 1)^{-1}}$

3. Prove that $(a + b)^{-1} \neq a^{-1} + b^{-1}$.

4. Factor the numerator of the result of Example 19.

⋆Answers to starred problems are to be found at the back of the book.

1.4 FRACTIONAL EXPONENTS AND RADICALS

The square root of a positive number x is the positive number $\sqrt{x}$ such that $\sqrt{x}\,\sqrt{x} = x$. In order to assign an exponent to x that corresponds to $\sqrt{x}$ and obeys the property $x^m \cdot x^n = x^{m+n}$, we define

$$\sqrt{x} = x^{1/2}$$

This is consistent because $\sqrt{x}\,\sqrt{x} = x$ and

$$x^{1/2} \cdot x^{1/2} = x^{1/2+1/2} = x^1 = x.$$

Similarly, for a cube root, $\sqrt[3]{x}\,\sqrt[3]{x}\,\sqrt[3]{x} = x$ leads to $\sqrt[3]{x} = x^{1/3}$. In general, the nth root of any number can be written as

$$\boxed{\sqrt[n]{x} = x^{1/n}}$$

Example 21. *Simplify the expression* $9^{1/2} + 64^{1/3}$.

$$9^{1/2} + 64^{1/3} = \sqrt{9} + \sqrt[3]{64}$$
$$= 3 + 4$$
$$= 7 \;\checkmark$$

Fractional-exponent notation, such as $\frac{2}{3}$, has two equivalent interpretations, each of which is consistent with the property of exponents, $(x^m)^n = x^{mn}$.

$$x^{2/3} = x^{(1/3)\cdot 2} = (x^{1/3})^2$$
$$x^{2/3} = x^{2\cdot(1/3)} = (x^2)^{1/3}$$

In general,

$$\boxed{x^{m/n} = (x^{1/n})^m = (x^m)^{1/n}}$$

For computations, the first form is generally easier to work with.

Example 22. *Simplify the expression* $64^{2/3} + 4^{7/2}$.

$$64^{2/3} + 4^{7/2} = (64^{1/3})^2 + (4^{1/2})^7$$

$$= (4)^2 + (2)^7$$
$$= 16 + 128$$
$$= 144 \quad \checkmark$$

Example 23. *Simplify* $16^{-5/4}$.

$$16^{-5/4} = \frac{1}{16^{5/4}} = \frac{1}{(16^{1/4})^5} = \frac{1}{(2)^5} = \frac{1}{32} \quad \checkmark$$

The properties of integer exponents also apply to any rational exponents. The property $(x \cdot y)^m = x^m \cdot y^m$ is useful in simplifying radicals. For example,

$$\sqrt{x \cdot y} = (x \cdot y)^{1/2} = x^{1/2} \cdot y^{1/2} = \sqrt{x} \sqrt{y}$$

Thus,

$$\boxed{\sqrt{x \cdot y} = \sqrt{x} \cdot \sqrt{y}} \quad \text{and similarly,} \quad \boxed{\sqrt{\frac{x}{y}} = \frac{\sqrt{x}}{\sqrt{y}}}$$

Example 24. *Simplify* $\sqrt{75}$.

$$\sqrt{75} = \sqrt{25 \cdot 3} = \sqrt{25} \sqrt{3} = 5 \sqrt{3} \quad \checkmark$$

The square root of a number can be simplified in this manner whenever the number can be written as a perfect square times another integer.

It is sometimes desirable to eliminate irrational numbers from denominators of fractions, especially when computations are forthcoming. It is simpler to divide a rational constant (often an integer) into an irrational number than to do the reverse of this process. For example,

$$\frac{7}{\sqrt{2}} \doteq \frac{7}{1.414214} \qquad \begin{cases} \text{The symbol} \doteq \text{means ap-} \\ \text{proximately equal to. Here} \\ 1.414214 \doteq \sqrt{2}. \end{cases}$$

is a more difficult division than is

$$\frac{7\sqrt{2}}{2} \doteq \frac{7(1.414214)}{2}$$

An irrational number can be removed from the denominator of a fraction by multiplying both numerator and denominator by an appropriate irrational number. Examples of such techniques are shown next.

Example 25. *Rationalize the denominator of* $\dfrac{2}{\sqrt{5}}$.

$$\frac{2}{\sqrt{5}} = \frac{2}{\sqrt{5}} \cdot \frac{\sqrt{5}}{\sqrt{5}} = \frac{2\sqrt{5}}{5} \quad \checkmark$$

Example 26. *Rationalize the denominator of* $\dfrac{2\sqrt{3}}{\sqrt{8}}$.

Although we could multiply both numerator and denominator by $\sqrt{8}$, it is simpler to multiply by $\sqrt{2}$, which is sufficient to produce a rational denominator.

$$\frac{2\sqrt{3}}{\sqrt{8}} = \frac{2\sqrt{3}}{\sqrt{8}} \cdot \frac{\sqrt{2}}{\sqrt{2}} = \frac{2\sqrt{6}}{4} = \frac{\sqrt{6}}{2} \quad \checkmark$$

Example 27. *Rationalize the denominator of* $\dfrac{5}{4 - \sqrt{3}}$.

$$\frac{5}{4 - \sqrt{3}} = \frac{5}{4 - \sqrt{3}} \cdot \frac{4 + \sqrt{3}}{4 + \sqrt{3}} = \frac{5(4 + \sqrt{3})}{16 - 3} = \frac{5(4 + \sqrt{3})}{13} \quad \checkmark$$

Do you see the application here of factoring the difference of two squares?

Here is an expression that is the result of applying a calculus operation called differentiation to the product $x^3(x^2 + 10)^{1/2}$.

$$x^4(x^2 + 10)^{-1/2} + 3x^2(x^2 + 10)^{1/2}$$

The problem now is to simplify this expression. Note the common factors: (1) the highest power of x common to both, namely, x^2; and (2) the lowest power of $(x^2 + 10)$ common to both terms, namely, $(x^2 + 10)^{-1/2}$.† The result of factoring out $x^2(x^2 + 10)^{-1/2}$ is

$$x^2(x^2 + 10)^{-1/2}[(x^2) + 3(x^2 + 10)]$$

or

$$x^2(x^2 + 10)^{-1/2}(x^2 + 3x^2 + 30)$$

or

$$x^2(x^2 + 10)^{-1/2}(4x^2 + 30)$$

A 2 can be factored from $4x^2 + 30$ to yield

$$2x^2(x^2 + 10)^{-1/2}(2x^2 + 15)$$

---

†When negative powers appear, factor out the *lowest* power common to all terms.

To eliminate negative exponents, the factor $(x^2 + 10)^{-1/2}$ can be written as $1/(x^2 + 10)^{1/2}$ or as $1/\sqrt{x^2 + 10}$. The result is

$$\frac{2x^2(2x^2 + 15)}{\sqrt{x^2 + 10}}$$

This is a most useful form, and the denominator is not rationalized. More on this in Problems 3 and 7, Exercise Set C.

Equations that contain square roots can often be solved by squaring both sides of the equation, that is, multiplying both sides by equal quantities—themselves. If one side of the equation contains the sum (or difference) of two radicals or the sum (or difference) of a radical and another number, then manipulation to isolate the radical should be carried out first, if possible. Otherwise, it will be necessary to square both sides twice.

If the radical equation contains cube roots, then both sides must be cubed in order to simplify the equation.

Example 28. *Solve for x:* $\sqrt{6x - 1} = 2\sqrt{x}$.

After squaring both sides,

$$6x - 1 = 4x$$

or

$$x = \tfrac{1}{2} \quad \checkmark$$

You should *always check* the solutions to such equations, since the squaring process may introduce extraneous solutions. Above, $\frac{1}{2}$ checks in the original equation. On the other hand, the equation

$$\sqrt{3x + 1} + \sqrt{x - 1} = 0$$

has no solution. Although the mechanical problem-solving techniques produce $x = -1$ as a "solution," -1 does not check in the original equation, and there is no solution.

EXERCISE SET C

1. Simplify each expression.
   ⋆(a) $144^{1/2}$
   ⋆(b) $100^{3/2}$
   ⋆(c) $(9 + 16)^{1/2}$
   ⋆(d) $5^0 + 4^{3/2}$
   (e) $32^{2/5}$
   (f) $(25^{1/2} + 27^{1/3})^{1/3}$

⋆Answers to starred problems are to be found at the back of the book.

(g) $3 \cdot 4^{3/2} + 25^{3/2}$

*(h) $5 \cdot 4^{1/2}$

*(i) $16^{-1/2}$

(j) $81^{-3/4}$

(k) $(125^{1/3} + 9^{1/2})^{-1/3}$

*(l) $\sqrt{20}$

(m) $\sqrt{72}$

(n) $\sqrt{98}$

*(o) $\sqrt{12x^6 y^4}$

*(p) $\sqrt[3]{24}$

(q) $(27x^{12} y^6)^{4/3}$

*(r) $\sqrt{64x^{18}}$

*(s) $\sqrt[5]{-32n^5}$

(t) $\sqrt[3]{-27x^3 y^6}$

(u) $7\sqrt{28}$

(v) $5\sqrt{75}$

*(w) $\sqrt{\dfrac{9}{4}}$

(x) $\sqrt{\dfrac{17}{16}}$

(y) $\sqrt{\dfrac{7}{21}}$

*(z) $\sqrt{\dfrac{1}{2}}$

2. Rationalize the denominator of each fraction.

*(a) $\dfrac{3}{\sqrt{7}}$

*(b) $\dfrac{5}{\sqrt{18}}$

(c) $\dfrac{2\sqrt{5}}{\sqrt{6}}$

*(d) $\dfrac{3}{1 - \sqrt{2}}$

(e) $\dfrac{5}{2 + \sqrt{3}}$

*(f) $\dfrac{5}{\sqrt{3} - \sqrt{2}}$

(g) $\dfrac{10}{\sqrt{7} + \sqrt{5}}$

*(h) $\dfrac{5}{\sqrt[3]{7}}$

(i) $\dfrac{\sqrt{5} + 2}{\sqrt{5} - 2}$

(j) $\dfrac{\sqrt{6} - \sqrt{5}}{\sqrt{6} + \sqrt{5}}$

*(k) $\dfrac{7}{\sqrt{7}}$

(l) $\dfrac{x}{\sqrt{x}}$

3. Factor each expression completely, and eliminate negative exponents, if any.

*(a) $2(x + 1)^{1/2} + x(x + 1)^{-1/2}$

(b) $3x^2(3x^2 + 2)^{-1/2} + x(3x^2 + 2)^{1/2}$

*(c) $9x^2(1 - x^2)^{1/3} - 2x^4(1 - x^2)^{-2/3}$

(d) $5(x^2 + 9)^{2/5} + 4x^2(x^2 + 9)^{-3/5}$

*(e) $10(x^2 - 3)^{3/5} + 12x^2(x^2 - 3)^{-2/5}$

*(f) $4x^3(x^2 - 1)^{-1/2} - 2x^5(x^2 - 1)^{-3/2}$

(g) $15(x^2 + 2)^{-2/5} - 12x(x^2 + 2)^{-7/5}$

4. Solve and check each equation.

*(a) $\sqrt{2x + 1} = 8$

*(b) $\sqrt{5x - 9} = 3\sqrt{x}$

(c) $\sqrt{4x - 3} = \sqrt{8 - 5x}$

*(d) $\sqrt{x - 6} + 3 = \sqrt{x + 9}$

(e) $\sqrt{2x + 3} + \sqrt{2x - 5} = 6$

*(f) $(2x - 3)^{1/2} - 9 = 0$

$\star$(g) $\sqrt{\sqrt{\sqrt{x+4}}} = 5$ (h) $(4x+1)^{1/5} = 2$

(i) $\sqrt[3]{3x+5} - 4 = 0$

$\star$(j) $\sqrt[3]{4x-6} = \sqrt[3]{2(4-3x)} + 5$

$\star$(k) $\sqrt{3\sqrt{2x-3}} = \sqrt{x+3}$

5. Show that

$$(x + y\sqrt{w})^{-1} = \frac{x - y\sqrt{w}}{x^2 - wy^2}$$

6. In some calculus settings, it is necessary to rationalize the numerator of a fraction in order to compute the "limit" of a fraction. (The concept of limit is introduced in Chapter 10.) Rationalize the numerator of the fraction

$$\frac{\sqrt{x+h} - \sqrt{x}}{h}$$

7. In problem 3, factoring and eliminating negative exponents yielded a fraction in each case. In calculus, it is often necessary to determine where expressions such as those of problem 3 are equal to zero and also where they are undefined. If the expression is written as a fraction, then it is zero when its numerator is zero and it is undefined when its denominator is zero. For each expression below, determine where it is zero and where it is undefined because its denominator is zero.

$\star$(a) $\dfrac{3x-1}{\sqrt{x^2-9}}$ $\star$(b) $\dfrac{\sqrt{x}}{x^2+1}$

$\star$(c) $\dfrac{7}{x^4-64}$ (d) $\dfrac{x^2-9x}{\sqrt{x(x+1)}}$

$\star$(e) $x(x-3)^{-1/3}(3x+2)^4$ $\star$(f) $x^{-1} + x^{-2}$

(g) $3(6x-1)^{1/2} + 9x(6x-1)^{-1/2}$ (h) $2x(x+6)^{-4} - 4x^2(x+6)^{-5}$

8. Write the expression $2x^{-1/3} - 2$ as one fraction without any negative exponents.

9. Prove that $\sqrt{a+b} \neq \sqrt{a} + \sqrt{b}$.

10. Rationalize the denominator of each fraction.

$\star$(a) $\dfrac{2}{\sqrt[3]{7}}$ (b) $\dfrac{13}{\sqrt[3]{10}}$

$\star$(c) $\dfrac{5}{\sqrt[3]{32}}$ (d) $\dfrac{7}{\sqrt[3]{25}}$

(e) $\dfrac{5}{1-\sqrt[3]{2}}$ (*Hint:* difference of cubes)

(f) $\dfrac{3}{2+\sqrt[3]{5}}$

1.5 QUADRATIC EQUATIONS

In elementary algebra you solved equations of the form $ax^2 + bx + c = 0$ by factoring the expression $ax^2 + bx + c$ and then setting each factor equal to zero. The examples below illustrate the solution process.

$$x^2 - 7x + 12 = 0$$
$$(x - 3)(x - 4) = 0$$

$x - 3 = 0$	$x - 4 = 0$
$x = 3$	$x = 4$

$$2x^2 + x - 6 = 0$$
$$(2x - 3)(x + 2) = 0$$

$2x - 3 = 0$	$x + 2 = 0$
$2x = 3$	$x = -2$
$x = \dfrac{3}{2}$	

Equations of the form $ax^2 + bx + c = 0$ are called *quadratic equations* (in x). When they cannot be solved readily by factoring, then a technique called *completing the square* can be used. The equation $x^2 + 6x + 2 = 0$ cannot be solved by factoring involving integers, so completing the square will be used. First, the equation is written as

$$x^2 + 6x = -2$$

The expression on the left side is nearly the square of $x + 3$. Specifically, $(x + 3)^2 = x^2 + 6x + 9$. The square of $x + 3$ is 9 more than $x^2 + 6x$. So if $x^2 + 6x$ is replaced by $(x + 3)^2$ on the left, then the left side is 9 more than it was. To compensate, 9 should be added to the right side. Thus,

$$x^2 + 6x = -2$$

becomes

$$(x + 3)^2 = -2 + 9$$

or

$$(x + 3)^2 = 7$$

which leads to

$$x + 3 = \pm\sqrt{7}$$

and then

$$x = -3 \pm\sqrt{7}$$

The symbol $\pm$ is read "plus or minus." There are two numbers that satisfy the equation: $-3 + \sqrt{7}$ and $-3 - \sqrt{7}$.

You can always complete a square of the form $x^2 + nx + \cdots$ by noting that the square is $\left(x + \dfrac{n}{2}\right)^2$ and working from there. For example,

$$x^2 + 8x \cdots \longrightarrow (x + 4)^2$$
$$m^2 - 2m \cdots \longrightarrow (m - 1)^2$$
$$y^2 + 3y \cdots \longrightarrow \left(y + \frac{3}{2}\right)^2$$

Although the process of completing the square will be needed in Chapter 6 to determine the center and radius of a circle, it can be avoided here by completing the square once and for all, for the equation $ax^2 + bx + c = 0$ and obtaining a quadratic formula in terms of a, b, and c. That formula can then be used to solve all quadratic equations that cannot be solved by factoring involving integers.

To begin, multiply both sides of $ax^2 + bx + c = 0$ by $1/a$ in order to obtain a coefficient of 1 for x^2:

$$x^2 + \frac{b}{a}x + \frac{c}{a} = 0$$

Next, add $-c/a$ to both sides of the equation in order to remove the constant term from the left side.

$$x^2 + \frac{b}{a}x = -\frac{c}{a}$$

The left side is nearly the square of x plus half of b/a. If it is written as $\left(x + \dfrac{b}{2a}\right)^2$, then $\left(\dfrac{b}{2a}\right)^2$ must be added to the right side. Thus,

$$\left(x + \frac{b}{2a}\right)^2 = \left(\frac{b}{2a}\right)^2 - \frac{c}{a}$$

or

$$\left(x + \frac{b}{2a}\right)^2 = \frac{b^2}{4a^2} - \frac{c}{a}$$

The fractions on the right can be combined.

$$\left(x + \frac{b}{2a}\right)^2 = \frac{b^2 - 4ac}{4a^2}$$

which leads to

$$x + \frac{b}{2a} = \pm\sqrt{\frac{b^2 - 4ac}{4a^2}}$$

or

$$x = -\frac{b}{2a} \pm \sqrt{\frac{b^2 - 4ac}{4a^2}}$$

The right side can be simplified, since $\sqrt{4a^2} = 2a$.

$$x = -\frac{b}{2a} \pm \frac{\sqrt{b^2 - 4ac}}{2a}$$

Finally,

$$x = \frac{-b \pm \sqrt{b^2 - 4ac}}{2a} \qquad a \neq 0$$

The method of completing the square is not recommended for use in solving quadratic equations in which a is not 1 or b is not an even number. In either of these cases, fractions are unavoidable. Use of the formula is much simpler. Memorize this formula, if you do not already know it.

Example 29. *Solve $x^2 - 7x + 4 = 0$.*

In $x^2 - 7x + 4 = 0$, a is 1, b is -7, and c is 4. Substituting these numbers into

$$x = \frac{-b \pm \sqrt{b^2 - 4ac}}{2a}$$

yields

$$x = \frac{-(-7) \pm \sqrt{(-7)^2 - 4(1)(4)}}{2(1)}$$

or

$$x = \frac{7 \pm \sqrt{33}}{2} \quad \checkmark$$

Example 30. *Solve $x^2 - 4x - 1 = 0$.*

Here a is 1, b is -4, and c is -1. Thus,

$$x = \frac{-(-4) \pm \sqrt{(-4)^2 - 4(1)(-1)}}{2(1)}$$

or

$$x = \frac{4 \pm \sqrt{20}}{2}$$

Note that $\sqrt{20}$ can be simplified to $2\sqrt{5}$, and that such simplification will help to reduce the fraction.

$$x = \frac{4 \pm 2\sqrt{5}}{2}$$

$$= \frac{2(2 \pm \sqrt{5})}{2}$$

$$= 2 \pm \sqrt{5} \quad \checkmark$$

Example 31. *Solve* $6x^2 + 5x = 4$.

Change $6x^2 + 5x = 4$ to $ax^2 + bx + c = 0$ form, as

$$6x^2 + 5x - 4 = 0$$

So $a = 6$, $b = 5$, and $c = -4$. Now

$$x = \frac{-5 \pm \sqrt{(5)^2 - 4(6)(-4)}}{2(6)}$$

$$= \frac{-5 \pm \sqrt{121}}{12}$$

$$= \frac{-5 \pm 11}{12}$$

At this point, since there is no square root remaining, pursue each case $+$ and $-$.

$$x = \frac{-5 + 11}{12} = \frac{6}{12} = \frac{1}{2} \quad \checkmark$$

$$x = \frac{-5 - 11}{12} = \frac{-16}{12} = -\frac{4}{3} \quad \checkmark$$

This equation could have been readily solved by factoring.

EXERCISE SET D

1. Solve each equation by factoring.
   - ★(a) $x^2 + 5x + 4 = 0$
   - ★(b) $x^2 - 7x + 12 = 0$
   - ★(c) $y^2 - 3y - 10 = 0$
   - (d) $x^2 - 8x + 16 = 0$
   - ★(e) $x^2 + 5x - 14 = 0$
   - ★(f) $t^2 - 9 = 0$
   - ★(g) $2m^2 + 7m + 6 = 0$
   - ★(h) $3x^2 + 13x - 10 = 0$
   - (i) $5x^2 + 14x + 8 = 0$
   - (j) $2x^2 - 19x + 35 = 0$

★Answers to starred problems are to be found at the back of the book.

2. Solve each quadratic equation by completing the square.

⋆(a) $x^2 + 6x + 5 = 0$ ⋆(b) $x^2 + 8x - 2 = 0$

⋆(c) $n^2 - 4n + 1 = 0$ ⋆(d) $m^2 - 2m - 1 = 0$

(e) $x^2 - 12x + 9 = 0$ ⋆(f) $y^2 + y - 1 = 0$

(g) $x^2 - 3x - 8 = 0$ (h) $t^2 + 7t - \dfrac{1}{2} = 0$

⋆(i) $2x^2 + 3x - 4 = 0$ (j) $3x^2 - 4x + 1 = 0$

3. Solve each quadratic equation by using the quadratic formula.

⋆(a) $x^2 + 4x + 3 = 0$ ⋆(b) $x^2 - 7x + 10 = 0$

⋆(c) $t^2 - 2t - 1 = 0$ (d) $m^2 - 12m + 9 = 0$

⋆(e) $x^2 + 3x - 7 = 0$ (f) $x^2 - 5x = 8$

⋆(g) $x^2 = 2x + 5$ ⋆(h) $2x^2 + 5x - 10 = 0$

(i) $3x^2 - 9x + 2 = 0$ (j) $-5x^2 + 3x + 1 = 0$

⋆**4.** Solve the equation

$$\frac{x+2}{x+3} = \frac{x+8}{3x+2}$$

1.6 INEQUALITIES

In elementary algebra you used the intuitive concepts of "less than" and "greater than" and the symbols $<$ for less than and $>$ for greater than. The statement "5 is greater than 2" is written symbolically as $5 > 2$. Similarly, $2 < 5$; 2 is less than 5. The notation can be formally defined as follows.

1. $a < b$ means that $b - a$ is a positive number.

2. Also, $a > b$ means the same as $b < a$; that is, $a - b$ is a positive number.

On a number line $a < b$ means that a is to the left of b, or b is to the right of a.

$a < b$: ─────────────┼────────┼─────────
                                    a b

All real numbers satisfy the *trichotomy property*; namely, if a and b are real numbers, then exactly one of the following is true: $a < b$, $a = b$, or $a > b$.

We list now, without proof, some properties of inequalities. The letters a, b, and c represent any real numbers unless further restricted within.

1. If $a < b$ and $b < c$, then $a < c$.
2. If $a < b$, then $a + c < b + c$.
3. (i) If $a > b$ and $c > 0$, then $ac > bc$.
   (ii) If $a < b$ and $c > 0$, then $ac < bc$.
4. (i) If $a > b$ and $c < 0$, then $ac < bc$.
   (ii) If $a < b$ and $c < 0$, then $ac > bc$.

Property 4 says that if both sides of an inequality are *multiplied* by a *negative* number, then the direction of the inequality is reversed. For example, if $5 > 2$, then $-3(5) < -3(2)$, since $-15 < -6$. Also, if $-2 < 11$, then $-5(-2) > -5(11)$, since $10 > -55$.

In the next series of examples, inequalities are solved by manipulating them into a form which shows what values of the variable will make the original inequality a true statement. Properties 1–4 are used to help solve inequalities.

Example 32. *Solve $2x + 5 < 9$.*

Add -5 to both sides to get the x term alone. This uses property 2.

$$\begin{array}{rcr} 2x + 5 < & & 9 \\ -5 & & -5 \\ \hline 2x & < & 4 \end{array}$$

Now multiply both sides by $\frac{1}{2}$ (or equivalently, divide by 2) to get a coefficient of 1 for x. This is property 3 ii.

$$x < 2 \quad \checkmark$$

Example 33. *Solve $5x + 2 - 9x \geq 17$.*

Combining like terms on the left side produces

$$-4x + 2 \geq 17$$

Next, add -2 to both sides to get the x term alone. This is property 2.

$$-4x \geq 15$$

Now multiply both sides by $-\frac{1}{4}$ to get a coefficient of 1 for x. This is property 4 i.

$$x \leq -\frac{15}{4} \quad \checkmark$$

$\begin{cases} \textit{Multiplication by } -\frac{1}{4}, \text{ a} \\ \textit{negative} \text{ number, reverses the} \\ \text{direction of the inequality;} \\ \geq \text{ becomes } \leq. \end{cases}$

Example 34. *Solve* $3(x + 1) - 7 < 8x + 9$.

Distribute the 3 to eliminate parentheses.

$$3x + 3 - 7 < 8x + 9$$

Combine the like terms on the left side.

$$3x - 4 < 8x + 9$$

To get the x term alone, first add $+4$ to both sides. This uses property 2.

$$3x < 8x + 13$$

Then add $-8x$ to both sides. This, again, uses property 2.

$$-5x < 13$$

Obtain a coefficient of 1 for x by using property 4 ii.

$$x > -\frac{13}{5}$$

$\begin{cases} \text{After multiplication of both} \\ \text{sides by } -\frac{1}{5} \text{ and consequent} \\ \text{change of } < \text{ to } >. \end{cases}$

Techniques for solving inequalities containing products and/or quotients are presented by means of a series of examples.

Example 35. *Solve* $(x + 1)(x + 6) > 0$.

We seek all values of x that will make the product $(x + 1)(x + 6)$ positive. There are two cases in which the product of two numbers in positive:

1. If both numbers are positive.

2. If both numbers are negative.

Case 1 (both positive): $(x + 1) > 0$ and $(x + 6) > 0$.

$$\text{If } x + 1 > 0, \text{ then } x > -1$$
$$\text{If } x + 6 > 0, \text{ then } x > -6$$

The solution for this case is all x such that $x > -1$ *and* $x > -6$. Both inequalities must be satisfied, so we select $x > -1$ because if $x > -1$, then

$x > -6$ as well. Observe the use of the real number line to illustrate the conclusion.

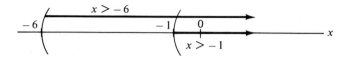

There are numbers greater than -6 (for example, -5) which are not greater than -1. Only if x is greater than -1 does it ensure that $x > -1$ *and* $x > -6$.

Case 2 (both negative): $(x + 1) < 0$ and $(x + 6) < 0$.

$$\text{If } x + 1 < 0, \text{ then } x < -1$$
$$\text{If } x + 6 < 0, \text{ then } x < -6$$

The solution for this case is all x such that $x < -1$ *and* $x < -6$. Both inequalities must be satisfied, so we select $x < -6$ because $x < -6$ satisfies $x < -1$ and $x < -6$.

There are numbers less than -1 (for example, -2) which are not less than -6. Only if x is less than -6 does it ensure that $x < -1$ *and* $x < -6$.

From case 1 we have $x > -1$. From case 2 we have $x < -6$. Thus, the entire solution is

$$x > -1 \quad \text{or} \quad x < -6 \checkmark$$

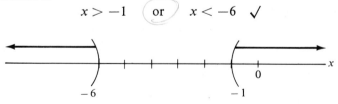

Notice the word "or." The two parts of the solution came from separate cases, and the cases $(+)(+)$, $(-)(-)$ could not occur simultaneously—thus, "or" rather than "and." Furthermore, you cannot have $x > -1$ *and* $x < -6$. Each one excludes the other. Thus, if $x > -1$, it cannot be true that the same x is also less than -6. Similarly, if $x < -6$, x cannot simultaneously be greater than -1.

The inequality $(x + 1)(x + 6) > 0$ of Example 35 might have been written originally in the multiplied form $x^2 + 7x + 6 > 0$, in which case factoring would have been the first step toward a solution.

Example 36. *Solve* $\dfrac{2}{x + 2} < 4$.

Multiply both sides of the inequality by $x + 2$, to eliminate the fraction and thus make it easier to solve. But wait! If you multiply by $x + 2$, are you multiplying by a positive number or a negative number? It could be either, so consider both cases.

Case 1: $\quad x + 2 \overset{+}{>} 0$ (which can also be written as $x > -2$).
If both sides of

$$\frac{2}{x + 2} < 4$$

are multiplied by a positive number $x + 2$, the result is

$$2 < (x + 2)4$$
$$2 < 4x + 8$$
$$-6 < 4x$$
$$-\frac{6}{4} < x$$

or

$$x > -\frac{3}{2}$$

Since this result, $x > -\frac{3}{2}$, occurs in the case where $x + 2 > 0$, that is, where $x > -2$, the solution for this case must satisfy both the inequalities

$$x > -\frac{3}{2} \quad \text{and} \quad x > -2$$

Note that $x > -\frac{3}{2}$ satisfies both, as shown by the overlap and shading of the common solution. Thus, $x > -\frac{3}{2}$ is the solution for this case.

Case 2: $\quad x + 2 \overset{-}{<} 0$ (or $x < -2$).

If both sides of

$$\frac{2}{x+2} < 4$$

are multiplied by a negative number $x + 2$, the result is

$$2 > (x + 2)4$$

since multiplication of both sides of an inequality by a negative number reverses the direction of the inequality. This then simplifies to

$$2 > 4x + 8$$
$$-6 > 4x$$
$$-\frac{6}{4} > x$$

or

$$x < -\frac{3}{2}$$

Since this result, $x < -\frac{3}{2}$, occurs in the case where $x + 2 < 0$, that is, where $x < -2$, the solution for this case must satisfy both the inequalities

$$x < -\frac{3}{2} \qquad \text{and} \qquad x < -2$$

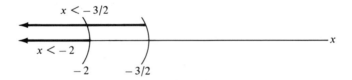

Notice that $x < -2$ satisfies both, as shown by the overlap and shading of the common solution. Thus, $x < -2$ is the solution for this case.

The solution for the problem is obtained by combining the solutions from Cases 1 and 2. Thus, the solution is

$$x > -\frac{3}{2} \qquad \text{or} \qquad x < -2 \quad \checkmark$$

Example 37. *Solve $x^2 - 3x - 10 \leq 0$.*

Since $x^2 - 3x - 10$ can be factored as $(x - 5)(x + 2)$, we proceed to solve $(x - 5)(x + 2) \leq 0$. Note that $(x - 5)(x + 2) = 0$ when $x = 5$ or $x = -2$. There are two cases for $(x - 5)(x + 2) < 0$.

Case 1 (negative)(positive): $x - 5 < 0$ and $x + 2 > 0$.

$$x - 5 < 0 \quad \text{and} \quad x + 2 > 0$$
$$x < 5 \quad \text{and} \quad x > -2$$
$$-2 < x < 5$$

Notice how $x < 5$ and $x > -2$ have been combined to indicate a solution of all numbers between -2 and 5, that is, all numbers that are greater than -2 and less than 5.

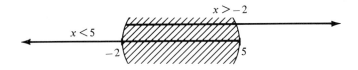

The overlap is where $x < 5$ *and* $x > -2$. Recall that $x = 5$ and $x = -2$ were also solutions of the inequality. Thus, the solution resulting from this case is $x \leq 5$ and $x \geq -2$, which can be combined as

$$-2 \leq x \leq 5$$

Case 2 (positive)(negative): $x - 5 > 0$ *and* $x + 2 < 0$.

$$x - 5 > 0 \quad \text{and} \quad x + 2 < 0$$
$$x > 5 \quad \text{and} \quad x < -2$$

It is impossible for a number to be both greater than 5 and less than -2. This case leads to no additional solutions.

In conclusion, the solution to $x^2 - 3x - 10 \leq 0$ is $-2 \leq x \leq 5$.

The same solution can also be obtained by graphing. If values are supplied for x to get values for the expression $x^2 - 3x - 10$, then points of the form $(x, x^2 - 3x - 10)$ are obtained. The graph of all such points is shown next. You can see that $x^2 - 3x - 10 \leq 0$ between -2 and 5.

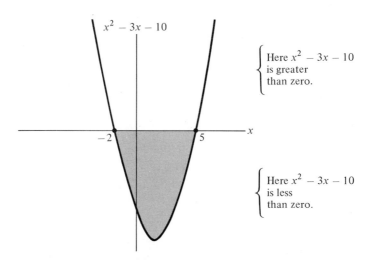

Example 38. *Solve* $\dfrac{x^2 - 7x + 12}{x + 5} < 0.$

The expression can be rewritten as

$$\frac{(x - 3)(x - 4)}{(x + 5)} < 0$$

There are several cases in which this expression could be less than zero (that is, negative).

1. $\dfrac{\text{(positive)(positive)}}{\text{(negative)}}.$

2. $\dfrac{\text{(negative)(negative)}}{\text{(negative)}}.$

3. $\dfrac{\text{(positive)(negative)}}{\text{(positive)}}.$

4. $\dfrac{\text{(negative)(positive)}}{\text{(positive)}}.$

Consider each of these cases.

1. $\dfrac{\text{(pos)(pos)}}{\text{(neg)}}$: $x - 3 > 0$ and $x - 4 > 0$ and $x + 5 < 0$

$\qquad\qquad\qquad\quad x > 3$ and $\qquad x > 4$ and $\qquad x < -5$

It is not possible for a number (x) to be greater than 3, greater than 4, *and* less than -5. There is no solution in this case.

2. $\dfrac{(\text{neg})(\text{neg})}{(\text{neg})}$: $x - 3 < 0$ and $x - 4 < 0$ and $x + 5 <$ 0
$x < 3$ and $x < 4$ and $x < -5$

The solution here is $x < -5$.

3. $\dfrac{(\text{pos})(\text{neg})}{(\text{pos})}$: $x - 3 > 0$ and $x - 4 < 0$ and $x + 5 >$ 0
$x > 3$ and $x < 4$ and $x > -5$

The solution here is $3 < x < 4$, as shown by the shaded overlap in the drawing.

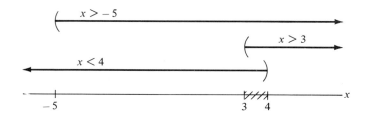

4. $\dfrac{(\text{neg})(\text{pos})}{(\text{pos})}$: $x - 3 < 0$ and $x - 4 > 0$ and $x + 5 >$ 0
$x < 3$ and $x > 4$ and $x > -5$

There is no solution in this case.

The solution to the problem is determined by combining the solutions of the cases that produced results: namely, cases 2 and 3. The solution is

$$x < -5 \quad \text{or} \quad 3 < x < 4 \quad \checkmark$$

Example 39. *Solve $x^3 + 2x^2 - 8x > 0$.*

The expression can be factored, as $x(x + 4)(x - 2) > 0$. There are four cases that produce positive products:

1. $(\text{pos})(\text{pos})(\text{pos})$.
2. $(\text{pos})(\text{neg})(\text{neg})$.
3. $(\text{neg})(\text{neg})(\text{pos})$.
4. $(\text{neg})(\text{pos})(\text{neg})$.

1. $(\text{pos})(\text{pos})(\text{pos})$: $x > 0$ and $x + 4 >$ 0 and $x - 2 > 0$
$x > 0$ and $x > -4$ and $x > 2$
Solution: $x > 2$

2. (pos)(neg)(neg): $x > 0$ and $x + 4 <$ 0 and $x - 2 < 0$
$x > 0$ and $x < -4$ and $x < 2$
Solution: none

3. (neg)(neg)(pos): $x < 0$ and $x + 4 <$ 0 and $x - 2 > 0$
$x < 0$ and $x < -4$ and $x > 2$
Solution: none

4. (neg)(pos)(neg): $x < 0$ and $x + 4 >$ 0 and $x - 2 < 0$
$x < 0$ and $x > -4$ and $x < 2$
Solution: $-4 < x < 0$

The solution is

$$x > 2 \quad \text{or} \quad -4 < x < 0 \ \checkmark$$

EXERCISE SET E

Solve each inequality.

*1. $x + 1 < 6$

2. $x - 3 \leq 10$

*3. $2x + 3 < 6$

4. $5x + 1 > 7$

5. $3x - 4 \leq -8$

6. $2x - 9 > -3$

*7. $3x + 1 - x \geq 13$

8. $5x + 3 + 2x < -4$

9. $2x - 7 - 9x \leq 6$

*10. $2(x + 3) - 6 \geq 9$

11. $3(x + 1) + 5 < 5$

*12. $4(x - 2) - 3 > x$

13. $2(3 + 2x) - 4 > 7x$

*14. $5x + 1 < 2(3x - 4) + 3$

15. $4x + 3 + 2x < 5 - 2x$

16. $4x + 2(3 + x) \geq 6x + 7$

*17. $(x - 5)(x + 2) \geq 0$

18. $(x + 4)(x + 7) > 0$

*19. $x^2 - 1 > 0$

*20. $x^2 - 10x + 16 < 0$

*21. $\dfrac{3}{x - 2} > 0$

22. $x^2 - 3x \leq 0$

23. $x^2 + 7x + 12 > 0$

*24. $\dfrac{4}{x + 1} < 0$

*25. $x^2 + 5x > 0$

*26. $x^2 - 7 < 0$

*27. $x^3 - 4x^2 + 3x < 0$

28. $x^3 + 3x^2 < 0$

*Answers to starred problems are to be found at the back of the book.

★29. $\dfrac{x}{x^2 - 9} > 0$ **★30.** $\dfrac{x + 5}{x - 3} < 0$

31. $\dfrac{x^2}{x - 7} < 0$ **★32.** $\dfrac{x^2 - 9x + 14}{x + 2} > 0$

★33. $\dfrac{x^2 + 2}{x^2 + 2x + 1} > 0$ **34.** $\dfrac{x + 4}{x^2 - 5x - 24} > 0$

1.7 ABSOLUTE VALUE

The *absolute value* (or *magnitude*) of a number, denoted $|x|$, is defined as

$$|x| = \begin{cases} x & \text{if } x \geq 0 \\ -x & \text{if } x < 0 \end{cases}$$

So the absolute value of any positive number is simply the number itself. The absolute value of zero is zero. The absolute value of any negative number is the opposite-signed number. Thus,

$$|+5| = 5$$
$$|\ 0| = 0$$
$$|-5| = -(-5) = 5$$

Consider the inequality $|x| < 5$. What numbers have absolute value less than 5? Obviously, any positive number less than 5 will satisfy this inequality. Zero also satisfies $|x| < 5$. Do all negative numbers satisfy $|x| < 5$? No, only negative numbers greater than -5. Numbers such as $-5, -6, -7$, etc., fail to satisfy $|x| < 5$ because their absolute values are not less than 5.

$$x = -5 \quad |x| = |-5| = 5$$
$$x = -6 \quad |x| = |-6| = 6$$
$$x = -7 \quad |x| = |-7| = 7$$

Thus, only numbers between -5 and 5 satisfy $|x| < 5$. The solution to $|x| < 5$ is $-5 < x < 5$.

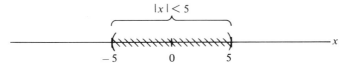

NOTE $x < 0$ *inside graph*

In general, for $a > 0$,

> If $|x| < a$,
> then $-a < x < a$.

Example 40. *Solve* $|x| < 9$.

$$|x| < 9 \quad \text{means} \quad -9 < x < 9 \quad \checkmark$$

Example 41. *Solve* $|2x| < 12$.

$$|2x| < 12 \quad \text{means} \quad -12 < 2x < 12$$

But we want x, not $2x$, so multiply by $\frac{1}{2}$.

$$-12 < 2x < 12$$

becomes

$$-6 < x < 6 \quad \checkmark$$

If you prefer, you can separate the inequality $-12 < 2x < 12$ into the two inequalities $-12 < 2x$ and $2x < 12$, get $-6 < x$ and $x < 6$, and put them back together as $-6 < x < 6$.

Example 42. *Solve* $|x + 9| < 15$.

$$|x + 9| < 15 \quad \text{means} \quad -15 < x + 9 < 15$$

Add -9 to all parts of the inequality to get x alone. This yields $-24 < x < 6$ as the solution.

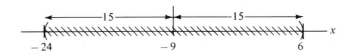

Note that this absolute-value inequality may be interpreted to mean that the distance between -9 and x must be less than 15 units on either side of -9.

Example 43. *Solve* $|5x - 8| < 2$.

$$|5x - 8| < 2 \quad \text{means} \quad -2 < 5x - 8 < 2$$

Add $+8$ to each part of the inequality to get

$$6 < 5x < 10$$

Now multiply by $\frac{1}{5}$ to get a coefficient of 1 for x.

$$\tfrac{6}{5} < x < 2 \;\; \checkmark$$

Consider the following computations.

$$|(+3)(+2)| = |+6| = 6 \qquad |+3|\cdot|+2| = 3\cdot2 = 6$$
$$|(+3)(-2)| = |-6| = 6 \qquad |+3|\cdot|-2| = 3\cdot2 = 6$$
$$|(-3)(+2)| = |-6| = 6 \qquad |-3|\cdot|+2| = 3\cdot2 = 6$$
$$|(-3)(-2)| = |+6| = 6 \qquad |-3|\cdot|-2| = 3\cdot2 = 6$$

These illustrate the relationship

$$|ab| = |a|\cdot|b|$$

The absolute value of the product is equal to the product of the absolute values.

The next two examples may appear somewhat strange. They have been taken from calculus settings, and you can be sure that you will see them again when you study calculus. The Greek letters epsilon (ϵ) and delta (δ) are used traditionally in such settings. Both letters, ϵ and δ, are used below as constants.

Example 44. *Show that if* $|5x - 15| < \epsilon$, *then* $|x - 3| < \dfrac{\epsilon}{5}$.

$$|5x - 15| < \epsilon$$
$$|5(x - 3)| < \epsilon$$
$$|5||x - 3| < \epsilon$$
$$5|x - 3| < \epsilon$$
$$|x - 3| < \frac{\epsilon}{5} \;\; \checkmark$$

Example 45. *Show that if* $|(2x - 3) - (-13)| < \epsilon$, *then* $|x + 5| < \delta$ *by finding a value for* δ *in terms of* ϵ.

$$|(2x - 3) - (-13)| < \epsilon$$
$$|2x - 3 + 13| < \epsilon$$
$$|2x + 10| < \epsilon$$
$$|2(x + 5)| < \epsilon$$

$$|2||x + 5| < \epsilon$$

$$2|x + 5| < \epsilon$$

$$|x + 5| < \frac{\epsilon}{2}$$

Since we want

$$|x + 5| < \delta$$

and we have

$$|x + 5| < \frac{\epsilon}{2}$$

choose

$$\delta = \frac{\epsilon}{2} \cdot \quad \checkmark$$

So far we have worked with absolute values that were *less than* some number; for example,

$$|x| < 9$$

$$|x + 9| < 15$$

$$|5x - 18| < 2$$

Now let's consider an example where the absolute value is *greater than* some number. Suppose that $|x| > 2$. Clearly, any number greater than 2 satisfies this. What negative numbers satisfy $|x| > 2$? Any number whose absolute value is greater than 2. For example, $-5, -12$, and $-31\frac{1}{2}$ satisfy $|x| > 2$. Any negative number less than -2 satisfies the inequality. Thus, $|x| > 2$ is satisfied by positive numbers greater than 2 and negative numbers less than -2.

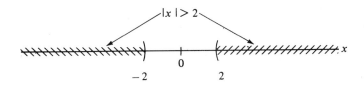

Thus, $|x| > 2$ means that

$$x > 2 \quad \text{or} \quad x < -2$$

Using the distance interpretation, the distance between 0 and x must always be greater than 2 units on either side of 0.

In general, for $a \geq 0$,

NOTE $\quad x \geq 0$

> If $|x| > a$,
>
> then $\quad x > a$
>
> or $\quad x < -a$

The inequalities $x > a$ or $x < -a$ are *not written* combined as $a < x < -a$ or as $-a > x > a$, because those forms suggest that x is greater than a and less than $-a$, and that is not possible for $a \geq 0$.

Example 46. *Solve* $|x| > 6$.

If $|x| > 6$, then

$$x > 6 \quad \text{or} \quad x < -6 \quad \checkmark$$

Example 47. *Solve* $|3x| > 27$.

$|3x| > 27$ means that

$$3x > 27 \quad \text{or} \quad 3x < -27$$
$$x > 9 \quad \text{or} \quad x < -9 \quad \checkmark$$

Example 48. *Solve* $|x - 8| > 6$.

$|x - 8| > 6$ means that

$$x - 8 > 6 \quad \text{or} \quad x - 8 < -6$$
$$x > 14 \quad \text{or} \quad x < 2 \quad \checkmark$$

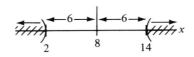

Example 49. *Solve* $|5x + 3| > 1$.

$|5x + 3| > 1$ means that

$$5x + 3 > 1 \quad \text{or} \quad 5x + 3 < -1$$
$$5x > -2 \quad \text{or} \quad 5x < -4$$
$$x > -\frac{2}{5} \quad \text{or} \quad x < -\frac{4}{5} \quad \checkmark$$

EXERCISE SET F

Solve each inequality in problems 1–28 below.

*1. $|x| < 6$ 2. $|x| < 19$

*3. $|4n| < 16$ 4. $|5t| < 20$

*5. $|7x| < 13$ 6. $|2x| < 11$

*7. $|m + 2| < 6$ 8. $|x - 17| < 6$

 9. $|x - 4| \leq 12$ *10. $|x + 5| \leq 0$

*11. $|3x + 5| < 10$ 12. $|4y - 2| < 10$

*13. $|4 - 7z| \leq 12$ *14. $|7 + 2x| < 15$

 15. $|8 - 9x| < 24$ 16. $|35 - x| < 10$

*17. $|3x| < \epsilon$ *18. $|x - 6| < \epsilon$

*19. $|x| > 7$ 20. $|x| > 65$

*21. $|5w| > 35$ 22. $|4x| > 86$

*23. $|x - 9| > 45$ 24. $|x + 16| \geq 6$

*25. $|2x - 7| > 16$ 26. $|4x - 3| > 30$

 27. $|5x + 9| > 20$ *28. $|2x + 12| > 5$

29. Show that if $|3x - 12| < \epsilon$, then $|x - 4| < \frac{\epsilon}{3}$.

30. Show that if $|6x - 18| < \epsilon$, then $|x - 3| < \frac{\epsilon}{6}$.

31. Show that if $|5x + 20| < \epsilon$, then $|x + 4| < \frac{\epsilon}{5}$.

32. Show that if $|(5x + 2) - 7| < \epsilon$, then $|x - 1| < \frac{\epsilon}{5}$.

33. Show that if $|(7x - 5) - 44| < \epsilon$, then $|x - 7| < \frac{\epsilon}{7}$.

34. Show that if $|(3x + 1) - (-11)| < \epsilon$, then $|x + 4| < \frac{\epsilon}{3}$.

35. Show that if $|(2x - 3) - (-5)| < \epsilon$, then $|x + 1| < \frac{\epsilon}{2}$.

36. Show that if $|\frac{1}{2}x - 2| < \epsilon$, then $|x - 4| < 2\epsilon$.

*Answers to starred problems are to be found at the back of the book.

★37. Find a δ such that if $|2x - 2| < .01$, then $|x - 1| < \delta$.

★38. Solve each inequality.

(a) $|7x + 2| < -1$

(b) $|5x - 3| > -2$

39. Is $\sqrt{x^2} = x$? Explain.

40. Prove that if x is a real number, then

$$\sqrt{x^2} = |x|$$

by considering all three possibilities for real numbers: positive, zero, and negative.

1.8 COMPLEX NUMBERS

The equation $x^2 + 1 = 0$ cannot be satisfied by substituting a real number for x, because if x is any real number, then $x^2 \geq 0$ and so $x^2 + 1 \geq 1$. Thus, $x^2 + 1 \neq 0$ for every *real* number. However, the numbers $+\sqrt{-1}$ and $-\sqrt{-1}$ satisfy this quadratic equation. Each checks in the original equation, $x^2 + 1 = 0$, because in each case $x^2 = -1$. The number $\sqrt{-1}$ is called i. It is an example of a so-called *imaginary number*. If $x^2 + 1 = 0$, then $x = \pm i$.

Here is a brief look at powers of i.

$$i^1 = \sqrt{-1}$$
$$i^2 = -1$$
$$i^3 = i^2 \cdot i = (-1)(i) = -i$$
$$i^4 = i^2 \cdot i^2 = (-1)(-1) = 1$$

It can be shown that

$$i = i^5 = i^9 = i^{13} = \cdots$$
$$i^2 = i^6 = i^{10} = i^{14} = \cdots$$
$$i^3 = i^7 = i^{11} = i^{15} = \cdots$$
$$i^4 = i^8 = i^{12} = i^{16} = \cdots$$

Square roots of negative integers other than -1 can be simplified to a multiple of i. The technique used is similar to that used earlier in the chapter to simplify $\sqrt{75}$ and others.

$$\sqrt{-4} = \sqrt{(4)(-1)}$$
$$= \sqrt{4}\sqrt{-1}$$
$$= 2 \cdot i \quad \text{or} \quad 2i \quad \checkmark$$

Similarly,

$$\sqrt{-7} = \sqrt{7}\sqrt{-1}$$
$$= \sqrt{7} \cdot i \quad \text{or} \quad i\sqrt{7} \quad \checkmark$$

Imaginary numbers are introduced when certain quadratic equations are solved by formula. For example, let's solve $x^2 + 2x + 3 = 0$. Here $a = 1, b = 2$, and $c = 3$. Thus,

$$x = \frac{-2 \pm \sqrt{4 - 12}}{2} = \frac{-2 \pm \sqrt{-8}}{2}$$

Since $\sqrt{-8}$ simplifies to $2i\sqrt{2}$, the preceding fraction can be simplified as follows:

$$x = \frac{-2 \pm 2i\sqrt{2}}{2} = \frac{2(-1 \pm i\sqrt{2})}{2} = -1 \pm i\sqrt{2} \quad \checkmark$$

The solutions are $-1 + i\sqrt{2}$ and $-1 - i\sqrt{2}$. Note that each of these numbers is a sum of two numbers—one real and one imaginary.

$$\underbrace{-1}_{\text{real part}} \quad \underbrace{+i\sqrt{2}}_{\substack{\text{imaginary} \\ \text{part}}}$$

This number is of the form $a + bi$, with a and b real numbers. Here $a = -1$ and $b = \sqrt{2}$. Numbers of the form $a + bi$, a and b real, are called *complex numbers.* Here are some other examples of complex numbers.

$$5 + i$$
$$7 - 3i$$
$$6i$$
$$12$$

The last two examples suggest that all real numbers are also complex numbers and that all imaginary numbers are also complex numbers. Both 12 and $6i$ are of the form $a + bi$. Specifically,

$$12 = 12 + 0i$$
$$6i = 0 + 6i$$

Two complex numbers $a + bi$ and $c + di$ are equal if $a = c$ and $b = d$. In other words, two complex numbers are equal if their real parts are equal and their imaginary parts are equal.

The sum of two complex numbers $a + bi$ and $c + di$ is the number $(a + c) + (b + d)i$. For example,

$$(15 + 3i) + (8 + 9i) = 23 + 12i$$

The difference of $a + bi$ and $c + di$ is the number $(a - c) + (b - d)i$. For example,

$$(15 + 3i) - (8 + 9i) = 7 - 6i$$

Let's multiply $(a + bi)$ by $(c + di)$ in order to determine an appropriate definition for multiplication of complex numbers.

$\left(\text{FOIL}\right)$
$$
\begin{aligned}
(a + bi)(c + di) &= a \cdot c + a \cdot di + bi \cdot c + bi \cdot di \quad (i^2) = -1 \\
&= ac + adi + bci + (-1)bd \\
&= (ac - bd) + (ad + bc)i
\end{aligned}
$$

The result is

$$(a + bi)(c + di) = (ac - bd) + (ad + bc)i$$

Unless you are doing a lot of multiplication with complex numbers, you might just as well multiply each as done in the preceding three-line process.

$$
\begin{aligned}
(3 + 4i)(5 - 2i) &= 3 \cdot 5 + 3(-2i) + 4i \cdot 5 + 4i(-2i) \quad (i^2) = -1 \\
&= 15 - 6i + 20i + 8 \quad (-8)(-1) = 8 \\
&= 23 + 14i \quad \checkmark
\end{aligned}
$$

The *complex conjugate* of the number $a + bi$ is the number $a - bi$. The product of a complex number and its conjugate is a nonnegative real number. Specifically,

$$
\begin{aligned}
(a + bi)(a - bi) &= a \cdot a + a(-bi) + bi(a) + bi(-bi) \\
&= a^2 - abi + abi + b^2 \\
&= a^2 + b^2
\end{aligned}
$$

As an example,

$$(2 + 3i)(2 - 3i) = 4 + 9 = 13 \quad \checkmark$$

Conjugates can be used to eliminate a complex number with nonzero imaginary part from the denominator of a fraction. In a sense, the process is division of complex numbers. For example, the division of $3i$ by $1 + i$ can be done by multiplying both numerator and denominator by the conjugate of $1 + i$. The result is a number of the form $a + bi$.

$$\frac{3i}{1+i} = \frac{3i}{1+i} \cdot \frac{1-i}{1-i} = \frac{3i+3}{2} \quad \text{or} \quad \frac{3}{2} + \frac{3}{2}i$$

In general,

$$\frac{a+bi}{c+di} = \frac{ac+bd}{c^2+d^2} + \frac{bc-ad}{c^2+d^2}i$$

EXERCISE SET G

1. Simplify each expression.

    ★(a) i^7 (b) i^{19} ★(c) i^{20}

    (d) i^{153} ★(e) $i^{23} + i^{27} + i^{31}$ (f) $3i^2 - 2i^4 + 6$

    ★(g) $\sqrt{-9}$ (h) $\sqrt{-16}$ (i) $\sqrt{-100}$

    ★(j) $\sqrt{-3}$ (k) $\sqrt{-5}$ ★(l) $\sqrt{-12}$

    (m) $\sqrt{-50}$ ★(n) $\sqrt{-243}$ (o) $\sqrt{-242}$

2. Solve each quadratic equation.

    ★(a) $x^2 + x + 1 = 0$ (b) $x^2 - x + 4 = 0$

    (c) $x^2 - 3x + 7 = 0$ ★(d) $x^2 - 2x + 5 = 0$

    ★(e) $x^2 + 3x + 9 = 0$ (f) $2x^2 + 7x + 12 = 0$

    ★(g) $3x^2 - 2x + 10 = 0$ (h) $-2x^2 + 3x - 4 = 0$

    ★(i) $x^2 + 8 = 0$ (j) $5x^2 + 18 = 0$

3. Perform each operation, and leave the result in $a + bi$ form.

    ★(a) $(2 + 3i) + (5 + i)$ (b) $(7 - 4i) + (1 + 6i)$

    ★(c) $(9 - i) - (-4 + 2i)$ (d) $(-6 + 3i) - (5 - 9i)$

    ★(e) $(8 - 3i) + (6)$ (f) $(7 + 3i) + (-5i)$

    ★(g) $(3 + 5i)(2 + 3i)$ (h) $(3 + 5i)(2 - 3i)$

    ★(i) $(8 + i)(8 + i)$ (j) $(i - 2)(3 - i)$

    (k) $(1 - i)^2$ ★(l) $(2 + i)^3$

★Answers to starred problems are to be found at the back of the book.

$\star$(m) $\dfrac{2}{1+i}$

(n) $\dfrac{-i}{1-2i}$

$\star$(o) $\dfrac{i-2}{3+i}$

(p) $\dfrac{-5+i}{2+3i}$

$\star$(q) $\dfrac{1+i}{2-i}$

(r) $\dfrac{5-4i}{5+4i}$

4. Prove that

$$\frac{a+bi}{c+di} = \frac{ac+bd}{c^2+d^2} + \frac{bc-ad}{c^2+d^2}i$$

chapter two

---

relations and functions

2.1 INTRODUCTION

Basic to the development of elementary calculus is the concept of function. Limits and the calculus operations of differentiation and integration are applied to functions. Consequently, it is necessary to be familiar with the definition and notation of functions. Part of each function is a domain and a range. Not only are domain and range essential in defining a function, but they are also very helpful in graphing a function.

The terms relation, function, domain, and range are defined and examined in this chapter. We begin with elementary set notation, since it is used to define the terms relation and function.

2.2 SET NOTATION

A *set* is a collection of things. The things which make up the set are called *elements* or *members* of the set. The elements of the set of integers between one and ten inclusive are 1, 2, 3, 4, 5, 6, 7, 8, 9, and 10. Such a set will be written as {1, 2, 3, 4, 5, 6, 7, 8, 9, 10}. Braces {   }

are used to specify a set. The elements of the set are placed inside the braces and are separated by commas. It does not matter in what order the elements are written. That is, the sets {1, 2, 3, 4, 5} and {3, 2, 1, 5, 4} are the same.

The symbol $\in$ is used to mean "is an element of" or "is in." For example, $5 \in \{2, 3, 5, 9, 17\}$ is read as 5 *is in* the set consisting of 2, 3, 5, 9, 17 or as 5 *is an element of* that set. The symbol $\notin$ means *is not an element of* or *is not in*. Thus the statement 7 *is not an element of* {1, 2, 3, 6} can be written $7 \notin \{1, 2, 3, 6\}$.

If every element of set A is also an element of set B, then A is called a *subset* of B, written $A \subseteq B$. The set {2, 3, 4, 6} is a subset of the set {1, 2, 3, 4, 5, 6}. It is also a subset of itself. In fact, $A \subseteq A$ for any set A.

The set containing no elements is called the *empty set* or *null set* and is denoted by $\varnothing$ or { }.

Sometimes three dots (. . .) are used to indicate a "continuation in the same manner." The notation is used to avoid writing all the members of the set explicitly. Thus, the set of whole numbers {1, 2, 3, 4, 5, 6, 7, 8, 9, 10, 11, 12, 13} can be written {1, 2, 3, 4, . . . , 13}. The set {14, 16, 18, 20, 22, 24, 26, 28, 30, 32, 34, 36} of even whole numbers between 14 and 36 inclusive can be written {14, 16, 18, . . . , 36}. The set of all whole numbers greater than 13 can be written as {14, 15, 16, 17, . . .}.

The set {2, 4, 6, 8, 10, . . .} is the set of all positive even integers. This can also be written as $\{x \mid x \text{ positive, even integer}\}$, which is read as "the set of all numbers x such that x is a positive, even integer." The vertical line $\mid$ is read as *such that*. This is particularly useful notation for some statements about sets. For example, how could we represent the set of *all* real numbers that are greater than zero (including all such rationals and irrationals)? Using the new notation we simply write this as $\{x \mid x \text{ real}, x > 0\}$, the set of all real numbers x such that x is greater than zero.

Example 1. *Indicate the set of all real numbers less than* 17, *except zero. Call the set* W.

$$W = \{m \mid m \text{ real}, m < 17, m \neq 0\}$$

Example 2. *Write the set of all odd numbers greater than* 7. *Call the set* R.

$$R = \{x \mid x \text{ odd}, x > 7\}$$

Example 3. *Represent the set of all real numbers that are between 3 and 5 exclusive, that is, all numbers that are both greater than 3 and less than 5. Call the set X.*

$$X = \{n \mid n \text{ real}, 3 < n < 5\}$$

EXERCISE SET A

1. Indicate which of the following statements are true and which are false.

⋆(a) $x \in \{a, b, c, x, y, z\}$ ⋆(b) $x \notin \{a, b, c, x, y, z\}$

⋆(c) $5 \in \{2, 4, 6, 8, 10\}$ ⋆(d) $5 \in \{1, 3, 15, 17, 19\}$

⋆(e) $d \in \{a, e, q, r\}$ ⋆(f) $* \notin \{\square, \triangle, *, \$\}$

(g) $-19 \in \{\text{integers}\}$ (h) $x \notin \{R, S, T, U, V\}$

(i) $341 \notin \{\text{odd numbers}\}$ (j) $5.6 \in \{\text{whole numbers}\}$

2. Use the notation of this section to represent the following sets.

⋆(a) all real numbers greater than 13

(b) all real numbers less than $\frac{1}{2}$

⋆(c) all real numbers not less than zero

⋆(d) the real numbers between -5 and 16 inclusive

(e) the odd numbers less than 10

(f) the even numbers between 3 and 95 exclusive

⋆(g) all even numbers between and including 4 and 300

⋆(h) the nonzero real numbers between but not including -50 and 400

(i) the integers between -600 and 700 inclusive

(j) all real numbers divisible by 2

⋆3. The set of all real numbers between 8 and 19 has been indicated as $\{x \mid x \text{ real}, 8 < x < 19\lvert$. Such an interval can also be represented by simpler *interval notation* as (8, 19). If the endpoints are to be included, the notation is [8, 19]. Also [8, 19) includes 8 and all real numbers between 8 and 19, with 19 excluded. Similarly, (8, 19] includes all the real numbers between 8 and 19, including 19 but not 8. The set [3, ∞) includes all reals that are 3 or larger. The set $(-\infty, 3]$ includes all the reals that are 3 or smaller. Represent each set of real numbers below by using interval notation.

(a) $\{x \mid 2 \leq x \leq 5\}$ (b) $\{x \mid -5 < x \leq 7\}$

(c) $\{t \mid 0 \leq t < 6\}$ (d) $\{m \mid -4 < m < 0\}$

(e) $\{y \mid y > 0\}$ (f) $\{x \mid x \leq -9\}$

⋆Answers to starred exercises are to be found at the back of the book.

2.3 RELATIONS

Consider the equation $y = 2x + 1$. For every real number x supplied, there is a corresponding real number y that makes the equation true. The pairs of x's and corresponding y's found by the equation $y = 2x + 1$ are an example of a _relation_ involving two sets of real numbers. Since the relation includes all real numbers x and y that satisfy the equation, it includes all ordered pairs that could be graphed by using the equation to determine points. Set notation can be used to express the relation as $\{(x, y)|y = 2x + 1\}$. Note that the pairs are _ordered pairs_, which means that the order in which the numbers are written is important. For example, the pair $(2, 5)$ is different from the pair $(5, 2)$.

We can think of a relation as a set of ordered pairs, a set of points. Accordingly, we can then define a _relation_ as a subset of the set of points that compose the xy plane.

2.4 FUNCTIONS

The concept of function involves three things: domain, range, and rule of correspondence. The _domain D_ is the set of all values that the first coordinate can take on. The domain is always a subset of the set of real numbers. The _range R_ is the set of all values that the second coordinate can take on. The range is always a subset of the set of real numbers. The _rule of correspondence_ specifies how to pair elements of the range with elements of the domain. Functions are usually named f, g, h, F, G, H, etc.

A _function f_ is a set of ordered pairs of real numbers (a relation) such that to each element of the domain there corresponds exactly one element of the range.

The relation $\{(x, y)|y = 2x + 1\}$ is an example of a function. In this example, the domain D is the set of all real numbers that can be used for x. The range R is the set of all real numbers that can be used for y. The rule of correspondence between the x's and the y's is as follows: Multiply x by 2 and then add 1 to the product to obtain y. A graph of this function shows that the domain is the set of all real numbers: $D = \{x|x \in \text{Reals}\}$. It is also apparent that $R = \{y|y \in \text{Reals}\}$; the range, too, is the set of all real numbers.

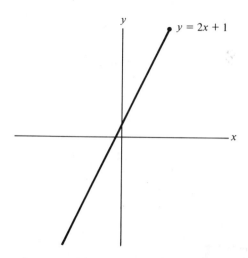

Techniques other than graphing are often used to determine the domain and range of a function, and these methods will be presented later. Throughout the book the following assumption is made: *Unless otherwise stated, the domain and range of a function (or relation) will be the domain and range implied by the rule of correspondence.* This is the largest possible domain and corresponding range. Whenever restrictions are intended, they will be stated explicitly.

If all relations were functions, there would be no need for defining both words. Not all relations are functions. The relation

$$\{(1, 3), (2, 5), (2, 4), (3, -7)\}$$

is not a function, because corresponding to the element 2 of the domain are two different elements (5 and 4) of the range.

Graphically, if any vertical line can be passed through two or more points of the relation, then the relation is not a function.

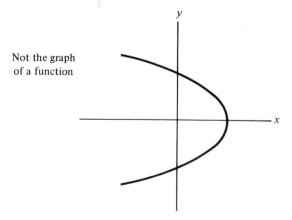

Not the graph
of a function

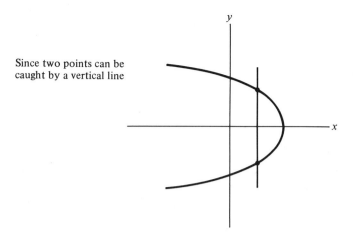

Since two points can be
caught by a vertical line

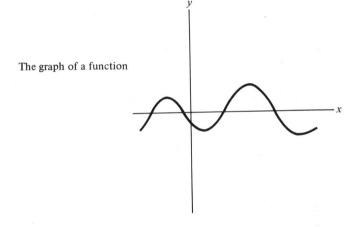

The graph of a function

Example 4. *Determine the domain and range of the function below.*

$$\{(1, 2), (2, 5), (3, 17), (5, -3)\}$$

The domain is $\{1, 2, 3, 5\}$, since these are the values that the first coordinate may assume.

The range is $\{2, 5, 17, -3\}$, since these are the values that the second coordinate takes on.

Example 5. *Find the implied domain and range of the function $y = 2x + 1$.*

Note that x can be any real number, since there is no real number we could substitute for x that would make the function undefined, for example, cause division by zero or the square root of a negative number. For every real number x there is a corresponding y. To find the range, solve for x in terms of y and see what y values, if any, do not produce a corresponding x.

$$y = 2x + 1$$

becomes

$$y - 1 = 2x$$

or

$$x = \frac{y - 1}{2}$$

Now you can see that any real number can be substituted for y in order to produce a corresponding x. So y can be any real number. In other words, the range is all the real numbers. Earlier we determined that the domain was all the real numbers.

Example 6. *Find the implied domain and range of the function* $y = x^2$.

For any real value of x supplied, we obtain a real value for y. So there is no restriction on x. The domain of the function is all the real numbers.

Since the y values are obtained by *squaring* the real x values, y can never be negative. And any nonnegative number (y) can be obtained by squaring the proper x. Thus, the range of the function is all real $y > 0$.

The domain and range can also be determined from the graph of $y = x^2$.

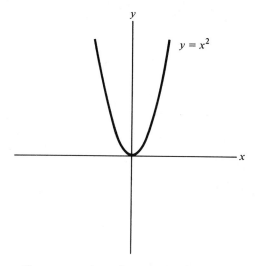

In the preceding examples, domain and range have been found three ways: studying the graph, algebraic manipulation, and examining the relation. When relations and functions are not written in a form that permits you to easily determine the domain and range by examination or by graph, some manipulation is useful. The following examples show various techniques for getting $y = \ldots$ form in order to determine the domain and $x = \ldots$ form in order to determine the range.

Example 7. *Determine the implied domain and range of $xy = 1$.*

We can obtain the $y =$ form by dividing both sides of $xy = 1$ by x. This yields

$$y = \frac{1}{x}$$

The relation is not defined for $x = 0$. Any other x value is acceptable. Thus, the domain is all the real numbers except 0; $D = \{x \mid x \text{ real}, x \neq 0\}$. Furthermore, you can now see that the relation is a function, because for any x value supplied there corresponds only one y value.

If the equation $xy = 1$ is solved for x by dividing both sides by y, the result is

$$x = \frac{1}{y}$$

So, similarly, the range is all real numbers y except $y \neq 0$; that is, $R = \{y \mid y \text{ real}, y \neq 0\}$.

Example 8. *Determine the implied domain and range of $y^2 - x = 0$.*

To find the range, solve for x by simply adding x to both sides of $y^2 - x = 0$. This produces

$$x = y^2$$

It is now clear that any real number can be substituted for y. Thus, the range is all the real numbers, or $R = \{y \mid y \in \text{Reals}\}$.

To find the domain, solve for y by first adding x to both sides of $y^2 - x = 0$. This yields

$$y^2 = x$$

Then

$$y = \pm\sqrt{x}$$

The domain includes all $x \geq 0$, or $D = \{x \mid x \text{ real}, x \geq 0\}$. The square root of any $x < 0$ is not a real number. Note that the relation is not a function. The $\pm$ indicates that for an x value supplied (say, $x = 9$), there are two y values ($y = \pm 3$).

Example 9. *Determine the implied domain and range of $x^2 y^2 + y^2 - 9 = 0$.*

To solve for y, first add 9 to both sides of the equation.

$$x^2 y^2 + y^2 = 9$$

Now factor out y^2 on the left side of the equation.

$$y^2(x^2 + 1) = 9$$

Divide both sides by $x^2 + 1$.

$$y^2 = \frac{9}{x^2 + 1}$$

$$y = \pm\sqrt{\frac{9}{x^2 + 1}}$$

There is no value of x which can be substituted in this form to cause division by zero or the square root of a negative number. So $D = \{x \mid x \in \text{Reals}\}$.

We can get x in terms of y by adding $-y^2 + 9$ to the original equation, $x^2 y^2 + y^2 - 9 = 0$. The result is

$$x^2 y^2 = 9 - y^2$$

Now divide both sides by y^2.

$$x^2 = \frac{9 - y^2}{y^2}$$

$$x = \pm\sqrt{\frac{9 - y^2}{y^2}}$$

Immediately you should see that y must not be zero, since division by zero is not defined, and so there is no x corresponding to a y value of 0. Furthermore, the fraction

$$\frac{9 - y^2}{y^2}$$

must not be negative, since the square root of a negative number is not a real number. Since y^2 is always positive, our only concern is that $9 - y^2$ not be negative. In other words, exclude from the range all values of y such that

$$9 - y^2 < 0$$

or

$$9 < y^2$$

or

$$y^2 > 9$$

If $y > 3$ or $y < -3$, then $y^2 > 9$. So the range is all y except $y = 0$, $y > 3$, $y < -3$. In other words,

$$R = \{y \mid y \text{ real}, -3 \leq y \leq 3, y \neq 0\}$$

or

$$R = \{y \mid y \text{ real}, |y| \leq 3, y \neq 0\}$$

Example 10. *Determine the implied domain and range of* $x^2y + xy + 7 = 0$.

First determine the domain by solving for y in terms of x.

$$x^2y + xy + 7 = 0$$
$$x^2y + xy = -7$$
$$y(x^2 + x) = -7$$
$$y = \frac{-7}{x^2 + x}$$

or

$$y = -\frac{7}{x(x + 1)}$$

If $x = 0$ or $x = -1$, then the fraction is not defined. All other x values have corresponding y values. The domain is therefore $\{x \mid x \text{ real}, x \neq 0, x \neq -1\}$.

Now solve for x in terms of y in order to determine the range. The original relation, $x^2y + xy + 7 = 0$, is a quadratic equation; that is, it is of the form $ax^2 + bx + c = 0$. It is quadratic in x. Note that it can be rewritten to make this more apparent.

$$yx^2 + yx + 7 = 0$$

where

$$a = y \quad \text{(the coefficient of } x^2\text{)}$$
$$b = y \quad \text{(the coefficient of } x\text{)}$$
$$c = 7 \quad \text{(the non-}x\text{ term)}$$

From

$$x = \frac{-b \pm \sqrt{b^2 - 4ac}}{2a}$$

we obtain

$$x = \frac{-y \pm \sqrt{y^2 - 4 \cdot y \cdot 7}}{2y}$$

or

$$x = \frac{-y \pm \sqrt{y^2 - 28y}}{2y}$$

If $y = 0$, we get the indeterminate form $0/0$. Returning to the original relation, $x^2y + xy + 7 = 0$, a value of 0 for y yields $7 = 0$. This is nonsense and indicates that there is no corresponding x for $y = 0$. Thus, 0 is not in the range of this relation.

The difference $y^2 - 28y$ cannot be negative or else we will have the square root of a negative number. So only y values such that $y^2 - 28y \geq 0$ can be in the range. The expression is equal to zero if $y = 0$ or $y = 28$; y can be 28, but it cannot be zero, as explained above. Now

$$y^2 - 28y > 0$$

if

$$y(y - 28) > 0$$

And there are two ways that this product can be positive: (positive)(positive) *or* (negative)(negative).

    Case 1:   (pos)(pos).

$$y > 0 \quad \text{and} \quad y - 28 > 0$$

or

$$y > 0 \quad \text{and} \quad y > 28$$

Thus,

$$y > 28 \quad \checkmark$$

    Case 2:   (neg)(neg).

$$y < 0 \quad \text{and} \quad y - 28 < 0$$

or

$$y < 0 \quad \text{and} \quad y < 28$$

Thus,

$$y < 0 \quad \checkmark$$

Therefore, the range is

$$\{y \mid y \text{ real}, y < 0 \text{ or } y \geq 28\} \quad \checkmark$$

Example 11. *Determine the implied domain and range of* $x^2 y^2 + 5y^2 + 2 = 0$.

First solve for y in terms of x in order to determine the domain.

$$x^2 y^2 + 5y^2 + 2 = 0$$
$$x^2 y^2 + 5y^2 = -2$$
$$y^2(x^2 + 5) = -2$$

$$y^2 = \frac{-2}{x^2 + 5}$$

$$y = \pm\sqrt{\frac{-2}{x^2 + 5}}$$

Since $x^2 \geq 0$ for all real numbers x, then $x^2 + 5 \geq 0$ for all real numbers x. This means that $\dfrac{-2}{x^2 + 5}$ is negative for all real numbers x. And since this fraction is under a radical indicating square root, there are no x values that produce real y values. The "domain" is the empty set. It follows that the range is also the empty set, since there are no y values. In fact, the equation $x^2 y^2 + 5y^2 + 2 = 0$ does not define a relation, since there are no ordered pairs.

Example 12. *Determine the implied domain of* $y = \sqrt{\dfrac{x + 3}{x - 8}}$.

Clearly, $x \neq 8$ or division by zero will be implied. Also, we insist that

$$\frac{x + 3}{x - 8} \geq 0$$

in order to avoid $\sqrt{\text{negative}}$. There are two cases which ensure that the fraction will be nonnegative.

Case 1: $\quad \dfrac{(\text{pos})}{(\text{pos})}.$

$$x + 3 > \quad 0 \quad \text{and} \quad x - 8 > 0$$
$$x > -3 \quad \text{and} \quad x > 8$$
$$\text{Conclude} \quad x > 8 \quad \checkmark$$

Case 2: $\quad \dfrac{(\text{neg})}{(\text{neg})}.$

$$x + 3 < \quad 0 \quad \text{and} \quad x - 8 < 0$$
$$x < -3 \quad \text{and} \quad x < 8$$
$$\text{Conclude} \quad x < -3 \quad \checkmark \quad \text{or} \quad x \leq -3$$

since $x = -3$ is acceptable because it makes the fraction equal to zero.

The domain is

$$D = \{x \mid x \text{ real}, x \leq -3 \text{ or } x > 8\} \quad \checkmark$$

EXERCISE SET B

★1. Determine which of the following are functions.

(a) $y = x + 3$ (b) $\{(1, 2)\}$

(c) $\{(0, 1), (1, 0)\}$ (d) $\{(1, 0), (1, 1)\}$

(e) $\{(1, 2), (3, 4), (-1, -2)\}$ (f) $y = x^2$

(g) $x^2 + y^2 = 25$ (h) $y^2 = 81 - x^2$

(i) $y = \pm x$ (j) $y = \dfrac{x + 8}{x - 2}$

For each of problems 2–37, determine the implied domain and range of each relation. Do not square both sides of any equation because that process often does not produce equivalent equations.

★2. $y = 8x$ **★3.** $y = \sqrt{x}$ (Note $\sqrt{x}$ means $+\sqrt{x}$.)

4. $y = -x$ **5.** $y = \pm x$

★6. $3xy = 7$ **★7.** $y = \dfrac{5}{x - 9}$

★8. $\{(3, 5), (5, 3), (6, -3)\}$ **★9.** $y = \sqrt{x - 3}$

10. $y = \sqrt{x^2 + 9}$ **11.** $y = x + 1$

★12. $y = x^2 + 1$ **13.** $y = \sqrt{x} + 1$

★14. $y^2 - 4x = 0$ **★15.** $x^2 + y^2 = 16$

16. $x^2 + y^2 + 9 = 0$ **17.** $x^3 - y = 0$

★18. $y = -\sqrt{x}$ **19.** $y = -\sqrt{x} + 2$

★20. $y = -\sqrt{x - 4}$ **21.** $x^2 y + y = 0$

★22. $x^2 y + xy - 4 = 0$ **★23.** $x^2 - y^2 = 16$

24. $4x^2 + 9y^2 = 36$ **25.** $y = -\sqrt{x^2 - 1}$

26. $x^2 y + 9y = 6$ **27.** $x^4 + y^4 - 25 = 0$

★28. $x^2 y^2 + y^2 - 13 = 0$ **★29.** $y = x^2 + 5x + 4$

30. $x^2 y^2 - 9x^2 - 4y^2 = 0$ **31.** $xy^2 - 2x - 16 = 0$

★32. $xy^2 + 3x - 7y = 0$ **33.** $xy^2 - 3x + 7y = 0$

★34. $x^2 y^2 + 3xy - 4 = 0$ **35.** $x^2 y^2 - 3y^2 = 7x$

36. $y = 3x^{1/3} - 9$ **37.** $y = 4 - x^{3/2}$

*Answers to starred problems are to be found at the back of the book.

2.5 FUNCTION NOTATION

A function f associates with each element of its domain exactly one element of its range. The notation $f(x)$ is frequently used to denote the second element of the ordered pair whose first element is x. Thus, ordered pairs take the form $(x, f(x))$. The notation $f(x)$ is read "f of x" or "f at x." It is the *value of f at x*. The correspondence can be seen as

$$x \xrightarrow{\ f\ } f(x)$$

If the function $y = 2x + 1$ is written as $f(x) = 2x + 1$, then

$$f(x) = 2(x) + 1$$

so for $x = 3$,

$$f(3) = 2(3) + 1 = 7$$

Symbolically,

$$3 \xrightarrow{\ f\ } 7$$

or

$$(3) \xrightarrow{\ f\ } 2(3) + 1 = 7$$

Note that $f(3)$ is the value of f at 3; here $f(3)$ means the same as "the y value in $y = 2x + 1$, when x is 3." Function notation is used extensively in calculus.

Example 13. *If* $f(x) = 3x^2 + 5x + 2$, *find* $f(0)$, $f(1)$, $f(-1)$, $f(x + 1)$, $f(r - 2)$.

$$f(x) = 3(x)^2 + 5(x) + 2$$
$$f(0) = 3(0)^2 + 5(0) + 2 = 2 \quad \checkmark$$
$$f(1) = 3(1)^2 + 5(1) + 2 = 10 \quad \checkmark$$
$$f(-1) = 3(-1)^2 + 5(-1) + 2 = 0 \quad \checkmark$$
$$f(x + 1) = 3(x + 1)^2 + 5(x + 1) + 2$$
$$= 3(x^2 + 2x + 1) + 5x + 5 + 2$$
$$= 3x^2 + 11x + 10 \quad \checkmark$$
$$f(r - 2) = 3(r - 2)^2 + 5(r - 2) + 2$$
$$= 3(r^2 - 4r + 4) + 5r - 10 + 2$$
$$= 3r^2 - 7r + 4 \quad \checkmark$$

Example 14. *Let* $f(x) = x^2 + 2x + 5$. *Let* Δx *represent a positive number.*
Find the value of

(a) $f(x + \Delta x)$

(b) $f(x + \Delta x) - f(x)$

(c) $\dfrac{f(x + \Delta x) - f(x)}{\Delta x}$

The symbol Δx represents one number and should be treated accordingly. It *does not mean* Δ times x. The symbol Δx and calculations like those below are used extensively in calculus courses.

(a) $f(x + \Delta x) = (x + \Delta x)^2 + 2(x + \Delta x) + 5$
$$= x^2 + 2x\,\Delta x + (\Delta x)^2 + 2x + 2\,\Delta x + 5 \quad \checkmark$$

(b) $f(x + \Delta x) \ominus f(x)$
$$= [x^2 + 2x\,\Delta x + (\Delta x)^2 + 2x + 2\,\Delta x + 5] \ominus (x^2 + 2x + 5)$$
$$= x^2 + 2x\,\Delta x + (\Delta x)^2 + 2x + 2\,\Delta x + 5 - x^2 - 2x - 5$$
$$= 2x\,\Delta x + (\Delta x)^2 + 2\,\Delta x \quad \checkmark$$

(c) $\dfrac{f(x + \Delta x) - f(x)}{\Delta x} = \dfrac{2x\,\Delta x + (\Delta x)^2 + 2\,\Delta x}{\Delta x}$
$$= \dfrac{\Delta x(2x + \Delta x + 2)}{\Delta x}$$
$$= 2x + \Delta x + 2 \quad \checkmark$$

Note that the division by Δx can be made because Δx is a positive number and therefore not equal to zero.

Example 15. *Given* $f(x) = x^2$ *and* $a = 3$, *find the value of*

$$\frac{f(x) - f(a)}{x - a} \qquad (x \neq a)$$

We must assume that $x \neq a$, that is, that $x \neq 3$. Otherwise, division by zero will be suggested.

$$\frac{f(x) - f(a)}{x - a} = \frac{(x^2) - (3^2)}{x - 3}$$
$$= \frac{x^2 - 9}{x - 3}$$
$$= \frac{(x + 3)(x - 3)}{(x - 3)}$$
$$= x + 3 \quad \checkmark$$

EXERCISE SET C

1. If $f(x) = 3x - 5$, find the value of

   (a) $f(1)$ ⋆(b) $f(2)$

   ⋆(c) $f(0)$ ⋆(d) $f(-5)$

   ⋆(e) $f(a)$ ⋆(f) $f(\tfrac{1}{2})$

   (g) $f(-\tfrac{1}{2})$ ⋆(h) $f(3m)$

   (i) $f(x + 1)$ ⋆(j) $f(x - 5)$

2. If $f(x) = x^2 + 8x - 1$, find the value of

   ⋆(a) $f(2)$ (b) $f(3)$

   (c) $f(10)$ ⋆(d) $f(0)$

   ⋆(e) $f(-1)$ (f) $f(-3)$

   ⋆(g) $f(t)$ ⋆(h) $f(\tfrac{1}{2})$

   ⋆(i) $f(-\tfrac{1}{4})$ ⋆(j) $f(x + 2)$

3. For each function below, compute and simplify the value of

$$\frac{f(x + \Delta x) - f(x)}{\Delta x} \qquad (\Delta x \neq 0)$$

   ⋆(a) $f(x) = x$ (b) $f(x) = 6x + 7$

   ⋆(c) $f(x) = x^2 - 4$ (d) $f(x) = 3x^2 + 5x - 1$

   ⋆(e) $f(x) = 12$ (f) $f(x) = -10x$

   ⋆(g) $f(x) = x^3$ (h) $f(x) = x^3 + 3x^2 + 4x - 7$

   ⋆(i) $f(x) = \dfrac{1}{x}$ (j) $f(x) = \dfrac{2}{x - 1}$

4. For each function, compute and simplify the value of

$$\frac{f(x) - f(a)}{x - a} \qquad (x \neq a)$$

   (a) $f(x) = x;\ a = 6$ (b) $f(x) = 3x + 4;\ a = 2$

   (c) $f(x) = x^2 - 3;\ a = 5$ (d) $f(x) = 2x^2 + 3x + 4;\ a = 5$

   (e) $f(x) = x^3;\ a = 4$ (f) $f(x) = 5x^2 - 8;\ a = 3$

⋆Answers to starred problems are to be found at the back of the book.

*5. A function is said to be *even* if $f(-x) = f(x)$ for all x in its domain. A function is *odd* if $f(-x) = -f(x)$ for all x in its domain. Determine which functions are even, which are odd, and which are neither.

(a) $f(x) = x$ (b) $f(x) = x^2$

(c) $f(x) = 5x + 2$ (d) $f(x) = 8$

(e) $f(x) = x^2 - x$ (f) $f(x) = x^3 + x$

(g) $f(x) = x^2 + 4$ (h) $f(x) = x^3 - 5$

(i) $f(x) = x^2 + x + 3$ (j) $f(x) = 3^x$

6. The function $s = f(t) = t^2 + 6t + 7$ gives the distance s in feet that a particle travels in t seconds. How far will the particle travel in 10 seconds? What is the meaning and value of $f(2)$?

7. Algebra of functions. Two functions can be added, subtracted, multiplied, and divided for all values in the domain common to both functions.

Addition: $(f + g)(x) = f(x) + g(x)$

Subtraction: $(f - g)(x) = f(x) - g(x)$

Multiplication: $(f \cdot g)(x) = f(x) \cdot g(x)$

Division: $(f \div g)(x) = \dfrac{f(x)}{g(x)}$ $(g(x) \neq 0)$

Determine $f + g, f - g, f \cdot g,$ and $f \div g$ for each function.

*(a) $f(x) = x, g(x) = 3$

(b) $f(x) = x^2, g(x) = x + 1$

(c) $f(x) = x + 2, g(x) = x^2 + 9$

2.6 SOME SPECIAL FUNCTIONS

This section presents graphs and brief discussions of several important kinds of functions. Later chapters are devoted to the study of still other kinds of functions.

Example 16. *Graph* $f(x) = |x|$.

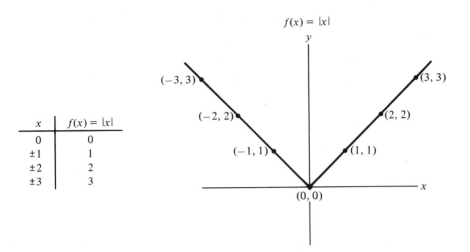

x	f(x) = |x|
0	0
±1	1
±2	2
±3	3

The domain of $f(x) = |x|$ is all the real numbers. The range is all non-negative real numbers; that is, $f(x) \geq 0$.

Example 17. *Graph* $f(x) = [\![x]\!]$.

The notation $[\![x]\!]$ means the greatest integer not exceeding x. Thus, $f(x) = [\![x]\!]$ is the *greatest integer function*. For each value of x supplied, the value of $f(x)$ obtained is the largest integer not exceeding x. This function is used extensively in computer programming, where truncation to the lower integer often occurs rather than rounding off to the nearer integer. A round-off can be obtained using the greatest integer function by specifying $[\![x + .5]\!]$.

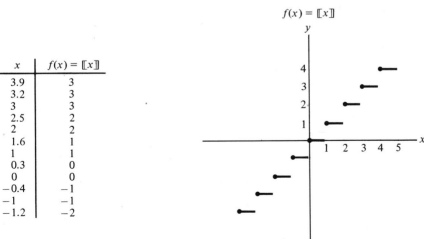

x	f(x) = [\![x]\!]
3.9	3
3.2	3
3	3
2.5	2
2	2
1.6	1
1	1
0.3	0
0	0
−0.4	−1
−1	−1
−1.2	−2

Note that $[\![-0.4]\!]$ is -1 not 0. Zero is greater than -0.4, whereas -1 is not greater than -0.4. Thus, -1 is the greatest integer in -0.4. Similarly, $[\![-1.2]\!]$ is -2 rather than -1.

The domain of $f(x) = [\![x]\!]$ is the set of all real numbers. The range is the set of integers.

Example 18. *Graph the function defined.*

$$f(x) = \begin{cases} 2x & x \geq 3 \\ 6 & 0 \leq x < 3 \end{cases}$$

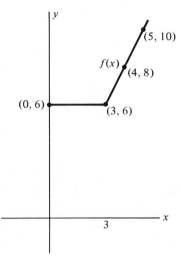

This is not really a special function, but the notation is different. For $x \geq 3$, the values of $f(x)$ are computed as $f(x) = 2x$. For $0 \leq x < 3$, $f(x)$ is 6. The function is only defined for $x \geq 0$. Thus, the domain is $x \geq 0$. The range is, from the graph, $f(x) \geq 6$.

Example 19. *Graph the function defined.*

$$f(x) = \begin{cases} -x - 2 & -2 \leq x \leq -1 \\ x & -1 < x \leq 1 \\ 2 - x & 1 < x \leq 2 \end{cases}$$

This function is not special either, although its shape is interesting. The domain is $-2 \leq x \leq 2$, since the function is only defined for x between -2 and 2. The range, from the graph, is $-1 \leq f(x) \leq 1$.

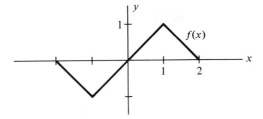

Example 20. *Graph* $g(x) = \dfrac{1}{x}$.

We have used $g(x)$ rather than $f(x)$ in this example to demonstrate that although f is the most frequently used letter, any letter can be used to name a function. The popular letters used for naming functions are $f, g, h, F, G,$ and H.

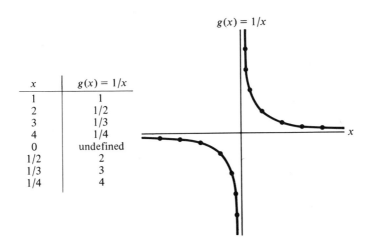

x	$g(x) = 1/x$
1	1
2	1/2
3	1/3
4	1/4
0	undefined
1/2	2
1/3	3
1/4	4

If x values of $-1, -2, -3, -4, -\frac{1}{2}, -\frac{1}{3},$ and $-\frac{1}{4}$ are supplied, then $g(x)$ will be $-1, -\frac{1}{2}, -\frac{1}{3}, -\frac{1}{4}, -2, -3,$ and -4, respectively.

Note that as the magnitude of x gets larger, the function gets closer to the line $y = 0$ (the x axis). However, the curve never actually reaches the line $y = 0$. We say that $y = 0$ is an *asymptote* of the curve $g(x) = 1/x$. Similarly, as the magnitude of $g(x)$ gets larger, the curve approaches the line $x = 0$ (the y axis). Thus, $x = 0$ is also an asymptote of $g(x) = 1/x$.

Example 21. *Graph* $F(x) = \dfrac{(x + 2)(x - 3)}{(x - 3)}$.

For all values of x except 3, the $x - 3$'s can be divided out to reduce

$F(x)$ to $x + 2$. When $x = 3$, the function is not defined. The graph is a line with a hole in it.

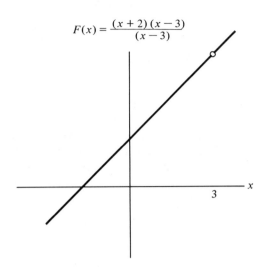

$$F(x) = \frac{(x + 2)(x - 3)}{(x - 3)}$$

If the function definition is extended to let $F(3) = 5$, as

$$F(x) = \begin{cases} \dfrac{(x + 2)(x - 3)}{(x - 3)} & (x \neq 3) \\ 5 & (x = 3) \end{cases}$$

then the curve will have no gaps; that is, it can be drawn entirely without lifting the pencil off the paper.

EXERCISE SET D

Graph each function and give its domain and range. Write the equations of asymptotes, if any.

★**1.** $f(x) = x^2$ ★**2.** $f(x) = 5$

★**3.** $f(x) = x^3$ **4.** $f(x) = \dfrac{1}{x^2}$

5. $g(x) = \dfrac{1}{x - 2}$ ★**6.** $f(x) = |x + 2|$

7. $h(x) = |x - 3|$ ★**8.** $f(x) = \sqrt{x}$

★Answers to starred problems are to be found at the back of the book.

★9. $f(x) = \begin{cases} 2 & x \geq 0 \\ x + 2 & x < 0 \end{cases}$

10. $f(x) = \begin{cases} x & x \geq 0 \\ -x & x < 0 \end{cases}$

★11. $h(x) = \begin{cases} -x & x \geq 0 \\ x & x < 0 \end{cases}$

12. $g(x) = \begin{cases} x^2 & x > 0 \\ 0 & x = 0 \\ -x^2 & x < 0 \end{cases}$

13. $F(x) = \begin{cases} [\![x]\!] & x \geq 0 \\ -[\![x]\!] & x < 0 \end{cases}$

★14. $G(x) = \begin{cases} |x| & x \geq 0 \\ -|x| & x < 0 \end{cases}$

★15. $f(x) = \dfrac{(x+5)(x+2)}{(x+2)}$

16. $f(x) = \dfrac{x^2 - 5x + 4}{x - 1}$

17. $f(x) = \dfrac{x^3}{x^2}$

★18. $f(x) = [\![\, |x| \,]\!]$

19. $g(x) = |[\![x]\!]|$

★20. $F(x) = -|x|$

21. $G(x) = \begin{cases} x & 0 \leq x < 1 \\ x - 1 & 1 \leq x < 2 \\ x - 2 & 2 \leq x < 3 \\ x - 3 & 3 \leq x < 4 \end{cases}$

22. $h(x) = \begin{cases} x + 2 & -2 \leq x \leq -1 \\ -x & -1 < x \leq 1 \\ x - 2 & 1 < x \leq 3 \\ 4 - x & 3 < x \leq 4 \end{cases}$

23. $F(x) = \begin{cases} 1 & x < 1 \\ 2 - x & 1 \leq x \leq 3 \\ -1 & 3 < x \leq 5 \\ x - 6 & 5 < x \leq 8 \\ 2 & x > 8 \end{cases}$

2.7 COMPOSITION OF FUNCTIONS

The *composite function* $(f \circ g)(x)$, read "f circle g of x," or "f composition g of x," or "f of g of x," is defined by

$$\boxed{(f \circ g)(x) = f(g(x))}$$

For example, if $f(x) = x + 5$ and $g(x) = \sqrt{x}$, then

$$(f \circ g)(x) = f(g(x))$$
$$= f(\sqrt{x})$$
$$= \sqrt{x} + 5 \quad \checkmark$$

and

$$(g \circ f)(x) = g(f(x))$$
$$= g(x + 5)$$
$$= \sqrt{x + 5} \quad \checkmark$$

Note that $(f \circ g)(x) \neq (g \circ f)(x)$ in this example. Occasionally, $(f \circ g)(x) = (g \circ f)(x)$, but usually not. In the next section, we shall see cases where $(f \circ g)(x) = (g \circ f)(x)$.

Let $f(x) = x^2 + 5x + 3$ and let $g(x) = 4x - 3$. Then

$$(f \circ g)(x) = f(g(x))$$
$$= f(4x - 3)$$
$$= (4x - 3)^2 + 5(4x - 3) + 3$$
$$= 16x^2 - 4x - 3 \quad \checkmark$$

and

$$(g \circ f)(x) = g(f(x))$$
$$= g(x^2 + 5x + 3)$$
$$= 4(x^2 + 5x + 3) - 3$$
$$= 4x^2 + 20x + 9 \quad \checkmark$$

Clearly, $(f \circ g)(x) \neq (g \circ f)(x)$ in this example.

The domain of $(f \circ g)(x)$ contains all numbers x in the domain of $g(x)$ for which $g(x)$ is in the domain of $f(x)$.

EXERCISE SET E

Compute $(f \circ g)(x)$ and $(g \circ f)(x)$ for each problem. Simplify whenever possible.

*1. $f(x) = x^2$, $g(x) = x + 3$

2. $f(x) = \sqrt{x}$, $g(x) = x - 1$

*3. $f(x) = \dfrac{1}{x}$, $g(x) = x^2$

4. $f(x) = x^2 + 2x + 5$, $g(x) = x + 2$

*5. $f(x) = \dfrac{x}{x + 1}$, $g(x) = 2 - x$

*Answers to starred problems are to be found at the back of the book.

*6. $f(x) = |x|, g(x) = 2x + 5$

7. $f(x) = [\![x + 1]\!], g(x) = x^2$

*8. $f(x) = \dfrac{3}{x + 2}, g(x) = \dfrac{1}{x}$

9. $f(x) = \dfrac{1}{x + 1}, g(x) = \dfrac{5}{x^2 - 3}$

*10. $f(x) = \dfrac{x + 2}{x - 2}, g(x) = \dfrac{1}{x}$

2.8 INVERSE FUNCTIONS

Recall that a relation is a set of ordered pairs. If we interchange the first and second coordinates of each ordered pair of a relation, we obtain the *inverse* of the relation. The domain of the relation becomes the range of the inverse relation. The range of the original relation is the domain of the inverse relation.

Example 22. *Let* $A = \{(1, 2), (3, 4), (5, 6)\}$.

If $A = \{(1, 2), (3, 4), (5, 6)\}$, then the domain of A is $\{1, 3, 5\}$ and the range of A is $\{2, 4, 6\}$.

The inverse of A is $A^{-1} = \{(2, 1), (4, 3), (6, 5)\}$. The domain of A^{-1} is $\{2, 4, 6\}$ and the range of A^{-1} is $\{1, 3, 5\}$.

If the relation is defined by an equation, then the inverse is readily obtained by interchanging the variables in the equation.

Example 23. *Find the inverse of* $y = 2x - 7$.   2 STEP

Relation:	$y = 2x - 7$	$\left\{ \begin{array}{l}\text{Interchange } x \text{ and } y \text{ to} \\ \text{obtain the inverse relation.}\end{array}\right.$
Inverse:	$x = 2y - 7$	

The inverse is usually manipulated to get it into the form $y =$. Thus, $x = 2y - 7$ becomes $2y = x + 7$ and then

$$y = \frac{x + 7}{2} \quad \checkmark$$

The graphs of a relation and its inverse are symmetric with respect to the line $y = x$. In other words, the line $y = x$ acts as a mirror with respect to a relation and its inverse. In view of this, an inverse relation can be sketched directly from the graph of the original relation by using the line $y = x$ as a mirror. Here is the graph of $y = 2x - 7$ and its inverse $y = \dfrac{x + 7}{2}$.

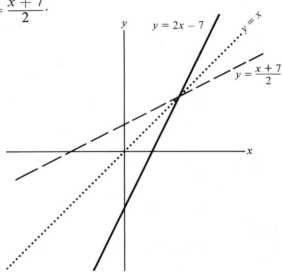

If the inverse of a relation satisfies the definition of function, then it is called the *inverse function* of the original relation. If no two distinct ordered pairs of a relation have the same second coordinate, then no two distinct ordered pairs of the inverse will have the same first coordinate. Thus, the inverse of such a relation is a function. The relation $y = x^2$ is a function.

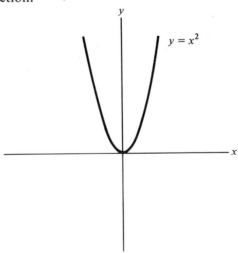

However, it has distinct ordered pairs having the same second coordinate. For example, $(2, 4)$ and $(-2, 4)$ are elements of the function $y = x^2$. So its inverse will not be a function. Its inverse is $y = \pm\sqrt{x}$, shown next.

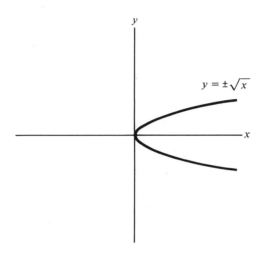

The implied domain D of $y = x^2$ includes all the real numbers. If that domain is restricted to either $D_1 = \{x \mid x \geq 0\}$ or $D_2 = \{x \mid x \leq 0\}$, then the inverse will be a function. Here are the two functions, their inverses, and their graphs.

$$f_1(x) = x^2 \qquad\qquad\qquad f_2(x) = x^2$$
$$D_1 = \{x \mid x \geq 0\} \qquad\qquad D_2 = \{x \mid x \leq 0\}$$
$$f_1^{-1}(x) = \sqrt{x} \qquad\qquad f_2^{-1}(x) = -\sqrt{x}$$

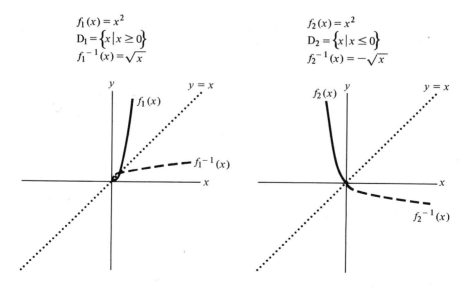

The inverse function of a function f is denoted f^{-1}. Furthermore,

$$(f \circ f^{-1})(x) = f(f^{-1}(x)) = x$$

for any function f and its inverse f^{-1}. Consider the functions in Example 23:

$$f(x) = 2x - 7 \quad \text{and} \quad f^{-1}(x) = \frac{x + 7}{2}$$

$$f(f^{-1}(x)) = f\left(\frac{x + 7}{2}\right)$$
$$= 2\left(\frac{x + 7}{2}\right) - 7$$
$$= x + 7 - 7$$
$$= x \quad \checkmark$$

EXERCISE SET F

Find the inverse of each of the relations in problems 1–19, and indicate whether the inverse is a function.

★**1.** $\{(0, 3), (1, 4), (2, 8), (-3, 6)\}$ ★**2.** $\{(1, 2), (1, 4), (1, 8)\}$

★**3.** $\{(3, 2), (4, -8), (6, -2), (10, 2)\}$ ★**4.** $y = 5x - 3$

5. $y = 9x + 2$ **6.** $y = \dfrac{2}{x}$

7. $y = x^2$ ★**8.** $y = x^2 - 3$

★**9.** $y = x^3$ **10.** $y = x^3 + 1$

11. $y = \dfrac{x + 5}{3x}$ **12.** $x^2 + y^2 = 25$

★**13.** $y = \dfrac{3}{x - 2}$ ★**14.** $y = \pm\sqrt{5 + x}$

15. $y = 4x^2 + x$ ★**16.** $xy^2 + x = 3$

★**17.** $y = x^2 + 3x + 5$ **18.** $x^2y + x = 3$

★Answers to starred problems are to be found at the back of the book.

19. $y = x^2 - 7x + 11$

20. Apply the result $f(f^{-1}(x)) = x$ as a check in all of problems 4–19 in which the inverse is a function.

21. Graph relations 4–9. Then sketch their inverses by using the fact that the line $y = x$ is a mirror with respect to a relation and its inverse.

2.9 SYMMETRY

SKIP

In the preceding section it was noted that a relation and its inverse are symmetric with respect to the line $y = x$. This made it very easy to sketch the graph of the inverse once the graph of the relation had been sketched. Recognition of certain kinds of symmetry can be used to simplify the graphing of some relations themselves. Here are some examples of symmetry and ways to discover it. The curve shown here, $y = x^2$, is symmetric with respect to the y axis.

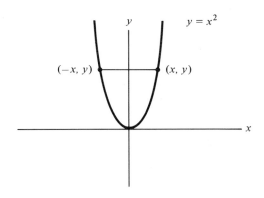

This means that the y axis is a mirror for the curve, reflecting what is on one side to the other side. For every point (x, y) on the curve there is a corresponding point $(-x, y)$. For example, corresponding to $(3, 9)$ is $(-3, 9)$. *Symmetry with respect to the y axis* can be tested for as follows. Substitute $-x$ for x in the original relation. If this new relation is the same as (or can be made the same as) the original, then the curve is symmetric with respect to the y axis.

The second curve shown, $y^2 - x^2 = 4$, is symmetric with respect to the x axis. This can be seen by looking at the illustration at the top of the next page.

$$y^2 - x^2 = 4$$

This means that the x axis is a mirror for the curve, reflecting what is on one side to the other side. For every point (x, y) on the curve, there is a corresponding point $(x, -y)$. For example, corresponding to $(0, 2)$ is $(0, -2)$. *Symmetry with respect to the x axis* can be tested for as follows. Substitute $-y$ for y in the original relation. If this new relation is the same as (or can be made the same as) the original, then the curve is symmetric with respect to the x axis. (You should note that $y^2 - x^2 = 4$ is also symmetric with respect to the y axis.)

The third curve shown, $y = 1/x$, is symmetric with respect to the origin.

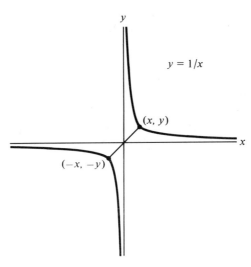

For every point (x, y) on the curve there is a corresponding point $(-x, -y)$. For example, corresponding to $(2, \frac{1}{2})$ is $(-2, -\frac{1}{2})$. *Symmetry with respect to the origin* can be tested for as follows. Substitute $-x$ for x and $-y$ for y in the original relation. If this new relation is the same as (or can be made the same as) the original, then the curve is symmetric with respect to the origin.

EXERCISE SET G

Test each relation for symmetry. Then graph each relation, using symmetry when available.

★**1.** $y = x^2 + 4$ ★**2.** $y^2 = x$

★**3.** $y = x^3$ **4.** $y = x^4$

★**5.** $y = |x|$ **6.** $y^2 - x^2 = 9$

7. $x^2 - y^2 = 1$ ★**8.** $y = -x$

★**9.** $x^2 + y^2 = 25$ **10.** $xy = 1$

2.10 TRANSLATION

Compare the graphs of

$$f(x) = x^2$$
$$g(x) = x^2 + 1$$
$$h(x) = x^2 - 2$$

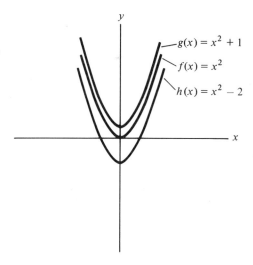

Point for point, the graph of $g(x) = x^2 + 1$ is one unit above the graph of $f(x) = x^2$. The graph of $h(x) = x^2 - 2$ is two units below $f(x) =$

★Answers to starred problems are to be found at the back of the book.

x^2. In general, the graph of $f(x) + c$ is c units above the graph of $f(x)$ if c is positive (or $-c$ units below $f(x)$, if c is negative).

Compare the graphs of

$$f(x) = x^2$$
$$g(x) = (x + 1)^2$$
$$h(x) = (x - 2)^2$$

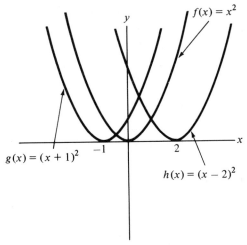

Point for point, the graph of $g(x) = (x + 1)^2$ is one unit to the left of the graph of $f(x) = x^2$. The graph of $h(x) = (x - 2)^2$ is two units to the right of $f(x) = x^2$. In general, the graph of $f(x + c)$ is c units to the left of the graph of $f(x)$ if c is positive (or $-c$ units to the right, if c is negative).

EXERCISE SET H

For each problem, sketch the graph of the function $f(x)$. Then graph $m(x)$, $n(x)$, $s(x)$, and $t(x)$ using the principles of vertical and horizontal translation suggested in this section.

1. $f(x) = x^2$.

$$m(x) = x^2 - 7 \qquad \text{↓ 7 UNITS}$$
$$n(x) = (x - 7)^2 \qquad \text{→ 7}$$
$$s(x) = x^2 + 5 \qquad \text{↑ 5}$$
$$t(x) = (x + 5)^2 \qquad \text{← 5}$$

2. $f(x) = |x|$.

$$m(x) = |x + 3| \qquad \text{<- 3 LEFT}$$
$$n(x) = |x| - 4 \qquad \text{↓ DOWN 4}$$
$$s(x) = |x| + 7 \qquad \text{↑ UP 7}$$
$$t(x) = |x - 1| \qquad \text{→ 1 RIGHT}$$

3. $f(x) = [\![x]\!]$.

$$m(x) = [\![x]\!] + 3$$
$$n(x) = [\![x + 3]\!]$$
$$s(x) = [\![x]\!] - 2$$
$$t(x) = [\![x - 2]\!]$$

4. $f(x) = \dfrac{1}{x}$.

$$m(x) = \frac{1}{x + 2}$$

$$n(x) = \frac{1}{x - 5}$$

$$s(x) = \frac{1}{x} + 4$$

$$t(x) = \frac{1}{x} - 3$$

2.11 GEOMETRIC PROBLEMS

SKIP

In this section, manipulative techniques of algebra are applied to problems involving geometry. The techniques include substitution, simplification of radicals, solution of quadratic equations, and equation manipulation. Many applied problems of differential and integral calculus involve settings and manipulations similar to those shown in the examples and exercises of this section.

Example 24. *The area of a circular region is given by $A = \pi r^2$. (a) Express the radius as a function of the area; (b) express the diameter of the circle as a function of the area.*

(a) To express the radius as a function of area, solve $A = \pi r^2$ for r.

$$A = \pi r^2$$

$$\frac{A}{\pi} = r^2 \qquad \text{(after dividing both sides by } \pi\text{)}$$

$$\sqrt{\frac{A}{\pi}} = r$$

or

$$r = f(A) = \sqrt{\frac{A}{\pi}} \quad \checkmark$$

The $\pm$ is omitted from in front of the radical because a radius cannot be negative. We conclude the $+$ case in this geometric setting.

(b) The diameter of a circle is twice the radius.

$$d = 2r$$

Since

$$r = \sqrt{\frac{A}{\pi}}$$

$$2r = 2\sqrt{\frac{A}{\pi}}$$

This means that

$$d = g(A) = 2\sqrt{\frac{A}{\pi}} \quad \checkmark$$

Example 25. *The volume contained by a cone is $V = \frac{1}{3}\pi r^2 h$, where r is the radius and h is the height. If the radius is equal to $\frac{2}{3}$ of the height, (a) determine the volume V as a function of h; (b) determine the volume as a function of r.*

(a) We have

$$V = \frac{1}{3}\pi r^2 h$$

and

$$r = \frac{2}{3}h$$

To get V as a function of h, eliminate r by substituting $\frac{2}{3}h$ for r.

$$V = \frac{1}{3}\pi r^2 h$$

becomes

$$V = \frac{1}{3}\pi (\tfrac{2}{3}h)^2 h \qquad \text{(after substituting } \tfrac{2}{3}h \text{ for } r)$$

Then

$$V = \frac{1}{3}\pi \tfrac{4}{25} h^2 h$$

Finally

$$V = f(h) = \frac{4\pi h^3}{75} \quad \checkmark$$

(b) To get V as a function of r, eliminate h. Since $r = \frac{2}{3}h$, then $h = \frac{3}{2}r$.
Thus,

$$V = \frac{1}{3}\pi r^2 h$$

becomes

$$V = \frac{1}{3}\pi r^2 (\tfrac{3}{2}r)$$

or

$$V = g(r) = \frac{5\pi r^3}{6} \quad \checkmark$$

Example 26. *Find the area of an equilateral triangle in terms of its side, s.*

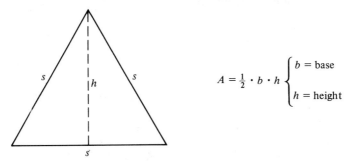

$$A = \tfrac{1}{2} \cdot b \cdot h \begin{cases} b = \text{base} \\ \\ h = \text{height} \end{cases}$$

The base is *s*, but we must find the height as a function of *s* in order to determine the area in terms of *s*. We shall now focus attention on half of the triangle.

By the Pythagorean Theorem,

$$s^2 = h^2 + \left(\frac{s}{2}\right)^2$$

Thus,

$$s^2 = h^2 + \frac{s^2}{4}$$

or

$$s^2 - \frac{s^2}{4} = h^2$$

The terms on the left can be combined by changing s^2 to a fraction with a denominator of 4.

$$\frac{4s^2}{4} - \frac{s^2}{4} = h^2$$

or

$$\frac{3s^2}{4} = h^2$$

Finally,

$$h = \sqrt{\frac{3s^2}{4}}$$

The term on the right can be simplified, if desired, by noting that $\sqrt{s^2} = s$
(because $s > 0$) and $\sqrt{4} = 2$.

$$h = \frac{s\sqrt{3}}{2}$$

Thus,

$$A = \frac{1}{2} \cdot b \cdot h$$

becomes

$$A = \frac{1}{2} \cdot s \cdot \frac{s\sqrt{3}}{2}$$

or

$$A = f(s) = \frac{s^2\sqrt{3}}{4} \quad \checkmark$$

Example 27. *Given the following triangle, write x as a function of y.*

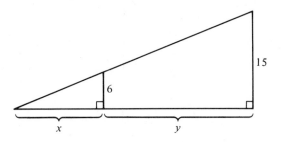

Note the following similar triangles that appear in the above triangle:

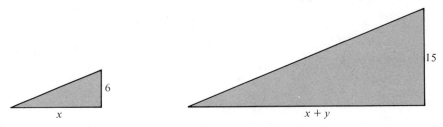

Similar triangles have proportional sides. This information can be used to set
up the proportion

$$\frac{x}{6} = \frac{x+y}{15}$$

which becomes

$$15x = 6(x + y)$$

Next proceed to multiply out the term on the right and then combine all x
terms on the same side.

$$15x = 6x + 6y$$

$$9x = 6y$$

$$x = \frac{6y}{9}$$

$$x = f(y) = \frac{2y}{3} \quad \checkmark$$

Example 28. *A piece of wire of length 23 units is to be cut into two pieces. One of the pieces will be bent into a square; the other will be bent into a circle. If the length of the piece used for the square is x, express the total area of the square and circular regions formed in terms of x.*

23

If one piece is of length x, then the other is of length $23 - x$.

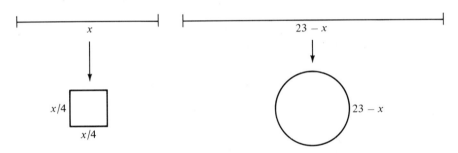

The square has sides of length $x/4$. Its area is any side squared.

$$A_\square = \left(\frac{x}{4}\right)^2 = \frac{x^2}{16}$$

The circle has a circumference of $23 - x$, where $C = 2\pi r$. To find the area of the circular region we need the radius, since $A = \pi r^2$. Thus,

$$C = 2\pi r$$

becomes

$$23 - x = 2\pi r \qquad \text{(since } C = 23 - x\text{)}$$

or

$$r = \frac{23 - x}{2\pi}$$

Now

$$A_\bigcirc = \pi r^2$$

$$= \pi \cdot \left(\frac{23 - x}{2\pi}\right)^2$$

$$= \frac{\pi(23 - x)^2}{4\pi^2}$$

$$= \frac{(23 - x)^2}{4\pi}$$

So the total area is

$$A_\square + A_\bigcirc = \frac{x^2}{16} + \frac{(23 - x)^2}{4\pi}$$

or

$$\text{area} = f(x) = \frac{x^2}{16} + \frac{(23 - x)^2}{4\pi} \quad \checkmark$$

Example 29. *Consider a right circular cone inscribed within a sphere of radius r. Express the height h of the cone in terms of the radius r of the sphere and the radius R of the cone.*

The sphere   The cone

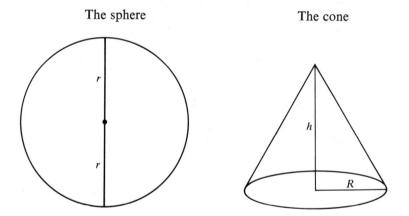

The cone inscribed within the sphere

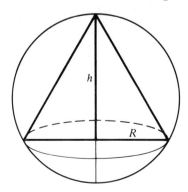

We can focus on

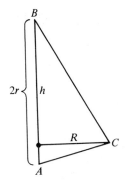

From geometry, the angle marked C is a right angle because it is inscribed in a semicircle. The two right triangles formed with common side R are similar, so their corresponding sides are proportional. One proportion that can be written is

$$\frac{h}{R} = \frac{R}{2r - h}$$

Multiply both sides of this equation by the common denominator $R(2r - h)$. The result is

$$R \cdot R = h(2r - h)$$
$$R^2 = 2rh - h^2$$
$$h^2 - 2rh + R^2 = 0$$

This is a quadratic equation in h, of form $ah^2 + bh + c = 0$, with $a = 1$, $b = -2r$, and $c = R^2$. Thus, by the quadratic formula,

$$h = \frac{-(-2r) \pm \sqrt{(-2r)^2 - 4(1)(R^2)}}{2(1)}$$
$$= \frac{2r \pm \sqrt{4r^2 - 4R^2}}{2}$$

$$= \frac{2r \pm \sqrt{4(r^2 - R^2)}}{2}$$

$$= \frac{2r \pm 2\sqrt{r^2 - R^2}}{2}$$

$$= \frac{2(r \pm \sqrt{r^2 - R^2})}{2}$$

Finally,

$$h = r \pm \sqrt{r^2 - R^2} \quad \checkmark$$

EXERCISE SET I

Solve each problem.

⋆1. Given $V = \frac{4}{3}\pi r^3$ for a sphere of radius r, express V as a function of diameter d rather than radius r.

⋆2. A cone has a radius equal to twice its height. Express the volume in terms of height, that is, without the radius.

⋆3. A rectangle has length x, width y, and area 25. Express the perimeter of the rectangle as a function of x.

4. Given a right triangle with legs a and b and hypotenuse 10, express the area as a function of b alone.

5. Express the area of an equilateral triangle in terms of its perimeter. Use A for area and P for perimeter.

⋆6. A man has a square sheet of paper 10 inches by 10 inches. Squares of size x by x are then cut from each corner of the sheet. The four flaps that are left after cutting are then folded up to form a box without a top. Express the volume of the box in terms of x.

⋆7. Express the area A of a circle as a function of its perimeter P.

8. Express the area A of a semicircle as a function of the perimeter, P, of the semicircle.

⋆Answers to starred problems are to be found at the back of the book.

★9. For the figure, express a in terms of b.

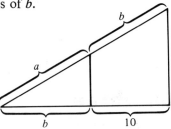

10. For the figure, express y in terms of x.

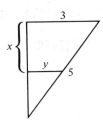

★11. A window in formed by placing a semicircular piece of glass above a rectangular piece. The perimeter of the entire window is 16. If the radius of the semicircular piece is r, express the area of the entire window in terms of r.

NOTE. The *Study Guide* contains a supplementary chapter on rational functions.

chapter three

---

exponential and
logarithmic functions

3.1 EXPONENTIAL FUNCTIONS

Any function of the form

$$f(x) = b^x \qquad (b > 0, b \neq 1, b \text{ constant})$$

is called an *exponential function*. What is different about such a function
is that the variable, x, is an exponent. The domain of any exponential
function is the set of all the real numbers; the exponent can be any real
number.† But $f(x) > 0$ only. The range contains only positive real
numbers. There is no way to raise a positive base to a power that produces a negative number or zero. For example,

$$2^3 = 8 > 0 \qquad \textit{ALL POSITIVE}$$
$$2^{1/2} = \sqrt{2} > 0$$

†Note that by permitting exponents to be arbitrary real numbers, irrational exponents must be considered as well as rational exponents. Irrational exponents are men-
tioned here for completeness, but they will not be discussed. Calculus can be used to explain
the nature of irrational exponents.

$$2^{-1} = \tfrac{1}{2} > 0 \qquad \text{ALL POSITIVE}$$
$$2^0 \ = 1 > 0$$

The graph of $f(x) = 2^x$ will serve as an example of the general features of exponential curves. Some points and a sketch are shown next.

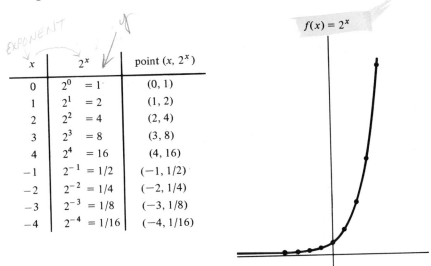

EXPONENT

x	2^x	point $(x, 2^x)$
0	$2^0 \ = 1$	$(0, 1)$
1	$2^1 \ = 2$	$(1, 2)$
2	$2^2 \ = 4$	$(2, 4)$
3	$2^3 \ = 8$	$(3, 8)$
4	$2^4 \ = 16$	$(4, 16)$
-1	$2^{-1} = 1/2$	$(-1, 1/2)$
-2	$2^{-2} = 1/4$	$(-2, 1/4)$
-3	$2^{-3} = 1/8$	$(-3, 1/8)$
-4	$2^{-4} = 1/16$	$(-4, 1/16)$

$f(x) = 2^x$

The x axis $(y = 0)$ is an asymptote of $f(x) = 2^x$. Also, the point $(0, 1)$ is on the graph of any exponential function $f(x) = b^x$.

You have probably heard comments about the exponential growth or rise of a corporation or population. Now you can see what is meant.

The section concludes with a sequence of examples that demonstrate some manipulations that can be applied in order to solve certain kinds of equations.

Example 1. *Solve for x:* $3^{4x-5} = 81$.

Write 81 as a power of 3: 3^4. Then

$$3^{4x-5} = 3^4$$

equal

Since the bases are equal, the exponents must also be the same. So

$$4x - 5 = 4 \qquad \Rightarrow \quad 4x = 4+5$$
$$4x = 9$$
$$x = \frac{9}{4} \quad \checkmark$$

Example 2. *Solve for x:* $(\frac{1}{2})^x = 64$.

Both $\frac{1}{2}$ and 64 are powers of 2, so express them as powers of 2 so that you can equate their exponents.

$$(\tfrac{1}{2})^x = (2^{-1})^x = 2^{-x}$$

and

$$64 = 2^6$$

Thus, the equation can be written as

$$2^{-x} = 2^6$$

from which we can conclude that

$$-x = 6$$

or

$$x = -6 \; \checkmark$$

EXERCISE SET A

1. Sketch the graph of each function.

(a) $f(x) = 3^x$ (b) $f(x) = 10^x$

(c) $f(x) = \left(\frac{1}{2}\right)^x$ (d) $f(x) = 2^{-x}$

(e) $f(x) = 3^{x+1}$ (f) $f(x) = 2^{1-x}$

2. Solve each of the following exponential equations.

⋆(a) $2^x = 16$ (b) $3^x = 243$

⋆(c) $2^x = \dfrac{1}{8}$ (d) $3^{-x} = 81$

⋆(e) $5^{2x} = 25$ ⋆(f) $4^{3x+1} = 256$

(g) $27^{x+2} = 64$ ⋆(h) $2^{5x} = 64^{x+1}$

(i) $3^{4-3x} = 27$ ⋆(j) $3^{-1} = 81^x$

(k) $7^{2x-1} = 1$ (l) $6^x = 36^{x+9}$

(m) $4^x = 2^{3x-1}$ ⋆(n) $49^{x+2} = 343$

⋆(o) $27^x = \left(\dfrac{1}{3}\right)^6$ (p) $7^{x+2} = 343$

(q) $4^{-x} = \dfrac{1}{32}$ ⋆(r) $\left(\dfrac{4}{3}\right)^{2x-3} = \dfrac{16}{9}$

(s) $\left(\dfrac{4}{3}\right)^x = \dfrac{9}{16}$ ⋆(t) $\left(\dfrac{2}{3}\right)^{x-5} = \left(\dfrac{9}{4}\right)^3$

⋆Answers to starred problems are to be found at the back of the book.

3. Exponential functions with bases of 0 and 1 were intentionally ignored. Graph and comment on the following two functions.

(a) $f(x) = 0^x$ (b) $f(x) = 1^x$

3.2 LOGARITHMS

Up to this point we have not learned a technique that would enable us to express the inverse of an exponential function. Think about this for a minute: How can you solve $y = b^x$ for x in terms of y? Before we answer that question, note that the graph of an exponential function shows that no two distinct ordered pairs have the same second coordinate. This means that the inverse of an exponential function will have no two distinct ordered pairs with the same first coordinate. In other words, the inverse of an exponential function is itself a function.

We now *define* the inverse of an exponential function to be a *logarithmic function*.

$$
\begin{array}{|c|}
\hline
\text{If } y = b^x \\
\text{then } x = \log_b y \\
\hline
\end{array}
\qquad
\begin{cases}
\text{In function notation,} \\
\text{if } f(x) = b^x, \text{ then} \\
f^{-1}(x) = \log_b x.
\end{cases}
$$

This is read as "x is the log(arithm) of y to the base b" or as "x is the logarithm to the base b of y." The logarithm is an exponent; it is the exponent to which b must be raised to produce y. Here are some numerical examples.

Exponential form	*Logarithmic form*
$8 = 2^3$	$\log_2 8 = 3$
$100 = 10^2$	$\log_{10} 100 = 2$
$3^4 = 81$	$\log_3 81 = 4$
$2^{-3} = \dfrac{1}{8}$	$\log_2 \dfrac{1}{8} = -3$
$7^0 = 1$	$\log_7 1 = 0$

Let us now write logarithmic functions in the form $y = \log_b x$ or as $f(x) = \log_b x$ and graph $f(x) = \log_2 x$ as an example. Note that we will select a value for $\log_2 x$ first, and then determine the corresponding x. It is much simpler this way. The domain of the function is all real numbers greater than zero; $D = \{x \mid x \text{ real}, x > 0\}$. Keep in mind that this means that logarithms of negative numbers are not defined. The range is all the real numbers; $R = \{y | y \in \text{Reals}\}$. The domain of the log function is the same as the range of the exponential function. The range of the log function is the same as the domain on the exponential function. The y axis $(x = 0)$ is an asymptote of $f(x) = \log_2 x$. Also, the point $(1, 0)$ is on the graph of any logarithmic function $f(x) = \log_b x$. The graph of $f(x) = \log_2 x$ and some of the points of the function are shown in the next figure.

A Logarithmic Function

the points   the graph

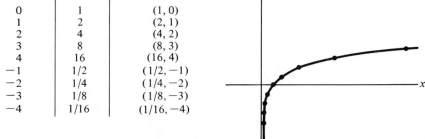

$\log_2 x$	x	point $(x, \log_2 x)$
0	1	$(1, 0)$
1	2	$(2, 1)$
2	4	$(4, 2)$
3	8	$(8, 3)$
4	16	$(16, 4)$
-1	1/2	$(1/2, -1)$
-2	1/4	$(1/4, -2)$
-3	1/8	$(1/8, -3)$
-4	1/16	$(1/16, -4)$

$f(x) = \log_2 x$

Since $f(x) = \log_2 x$ is the inverse of $f(x) = 2^x$, its graph could have been sketched from the graph of $f(x) = 2^x$ by using the line $y = x$ as a mirror. This technique was introduced in Chapter 2. The following figure shows both graphs.

The Log Function as the Inverse of the Exponential Function

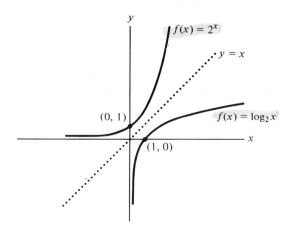

EXERCISE SET B

1. Write each exponential equation as an equivalent logarithmic equation.

⋆(a) $3^4 = 81$ ⋆(b) $25 = 5^2$

(c) $10^3 = 1000$ (d) $128 = 2^7$

⋆(e) $9^{1/2} = 3$ (f) $16^{1/4} = 2$

⋆(g) $4^{-1} = \dfrac{1}{4}$ (h) $5^0 = 1$

⋆(i) $2^a = b$ (j) $x^y = z$

2. Write each logarithm equation as an exponential equation.

⋆(a) $\log_{10} 100 = 2$ (b) $\log_2 64 = 6$

⋆(c) $\log_{25} 5 = \dfrac{1}{2}$ (d) $\log_9 1 = 0$

⋆(e) $\log_3 \dfrac{1}{9} = -2$ (f) $\log_d e = f$

3. Determine the value of each logarithm.

⋆(a) $\log_{10} 10{,}000 =$ (b) $\log_5 125 =$

⋆(c) $\log_2 1024 =$ (d) $\log_3 243 =$

⋆(e) $\log_{27} 3 =$ ⋆(f) $\log_{32} 2 =$

(g) $\log_{16} 8 =$ ⋆(h) $\log_{1/2} 4 =$

⋆(i) $\log_9 9 =$ (j) $\log_9 1 =$

⋆(k) $\log_9 9^2 =$ ⋆(l) $\log_b b^x =$

4. Graph each function and its inverse.

(a) $f(x) = \log_3 x$ (b) $f(x) = \log_{10} x$

(c) $f(x) = \log_{1/2} x$ (d) $g(x) = \log_{3/2} x$

(e) $h(x) = \log_2 (x + 1)$ (f) $F(x) = \log_2 (-x)$

3.3 PROPERTIES OF LOGARITHMS

There are three important properties of logarithms that make them useful for simplifying calculations. The properties are stated now, proved, and then applied afterward. They are stated for logarithms using any base $b > 0, \neq 1$.

⋆Answers to starred problems are to be found at the back of the book.

$$\boxed{\begin{aligned} \log_b M \cdot N &= \log_b M + \log_b N \\ \log_b \frac{M}{N} &= \log_b M - \log_b N \\ \log_b M^p &= p \cdot \log_b M \end{aligned}}$$

$\log_b M \cdot N = \log_b M + \log_b N$

Let $x = \log_b M$ and $y = \log_b N$. From $x = \log_b M$ we get the exponential form $M = b^x$. Similarly, from $y = \log_b N$ we have $N = b^y$. Thus,

$$M \cdot N = b^x \cdot b^y = b^{x+y}$$

so

$$\log_b M \cdot N = \log_b b^{x+y}$$

$$= x + y \qquad \begin{cases} \text{using the result of Exercise} \\ \text{Set B, problem 3(1)} \end{cases}$$

$$= \log_b M + \log_b N \qquad \begin{cases} \text{since } x = \log_b M \\ \text{and } y = \log_b N \end{cases}$$

$\log_b \frac{M}{N} = \log_b M - \log_b N$

Let $x = \log_b M$ and $y = \log_b N$. Then $M = b^x$ and $N = b^y$. Therefore

$$\frac{M}{N} = \frac{b^x}{b^y} = b^{x-y}$$

and

$$\log_b \frac{M}{N} = \log_b b^{x-y}$$

$$= x - y$$

$$= \log_b M - \log_b N$$

$\log_b M^p = p \cdot \log_b M$

Let $x = \log_b M$. Then $M = b^x$. Therefore

$$M^p = (b^x)^p = b^{px}$$

Thus,

$$\log_b M^p = \log_b b^{px}$$
$$= px$$
$$= p \cdot \log_b M$$

Example 3. *Compute the approximate value of* $\log_b \sqrt{\dfrac{6}{5}}$, *if* $\log_b 2 \doteq 0.3010$, $\log_b 3 \doteq 0.4771$, *and* $\log_b 5 \doteq 0.6990$.

The properties of logarithms yield, in steps,

$$\log_b \sqrt{\frac{6}{5}} = \log_b \left(\frac{6}{5}\right)^{1/2} \qquad \text{(notation)}$$

$$= \tfrac{1}{2} \log_b \frac{6}{5} \qquad \text{(third property of logarithms)}$$

$$= \tfrac{1}{2}(\log_b 6 - \log_b 5) \qquad \begin{cases}\text{second property of} \\ \text{logarithms}\end{cases}$$

$$= \tfrac{1}{2}(\log_b 3 \cdot 2 - \log_b 5) \qquad (\text{since } 6 = 3 \cdot 2)$$

$$= \tfrac{1}{2}(\log_b 3 + \log_b 2 - \log_b 5) \qquad (\text{first property of logarithms})$$

$$\doteq \tfrac{1}{2}(0.4771 + 0.3010 - 0.6990) \qquad \begin{cases}\text{using the values of the} \\ \text{logarithms}\end{cases}$$

$$\doteq \tfrac{1}{2}(0.0791)$$

$$\doteq 0.03955 \quad \checkmark$$

Example 4. *Write* $\log_b 5 + 2 \log_b 7$ *as the logarithm of a single number.*

$$\log_b 5 + 2 \log_b 7 = \log_b 5 + \log_b 7^2 \qquad \text{(third property of logarithms)}$$
$$= \log_b 5 + \log_b 49 \qquad (\text{since } 7^2 = 49)$$
$$= \log_b 5 \cdot 49 \qquad (\text{first property of logarithms})$$
$$= \log_b 245 \quad \checkmark \qquad (\text{since } 5 \cdot 49 = 245)$$

EXERCISE SET C

1. Let $\log_b 2 \doteq 5$, $\log_b 3 \doteq 7$, and $\log_b 5 \doteq 10$. Compute the approximate value of each number.

*(a) $\log_b 6$ (b) $\log_b 10$

(c) $\log_b 15$ *(d) $\log_b 4$

(e) $\log_b 27$ *(f) $\log_b 50$

(g) $\log_b 30$ *(h) $\log_b 1.5$

*Answers to starred problems are to be found at the back of the book.

*(i) $\log_b 0.6$

*(j) $\log_b \dfrac{2}{3}$

*(k) $\log_b \sqrt{3}$

(l) $\log_b \sqrt[4]{10}$

2. Write each expression as the logarithm of a single number.

*(a) $\log_r 7 + \log_r 5$

(b) $\log_r 4 + \log_r 3 + \log_r 10$

*(c) $\log_b 6 + 3 \cdot \log_b 5$

*(d) $\log_a 1 + \log_a 7 - \log_a 2$

*(e) $2 \log_b 6 + 2 \log_b 4$

(f) $\frac{1}{2} \log_b 9 + \frac{1}{4} \log_b 16 + \log_b 5$

3. Simplify each expression by using the properties of logarithms.

*(a) $\log_b 7 + \log_b x$

*(b) $\log_b x + \log_b x^2$

(c) $\log_b x^3 - \log_b x$

(d) $\log_b(x^2 - 1) - \log_b(x + 1)$

4. Write the expression below as the logarithm of one expression, that is, one logarithm rather than two logarithms. It can be done, but some insight is needed here because the bases of the two logarithms are different.

$$\log_x 4 + \log_t t \qquad (t \neq x)$$

5. What is wrong with the following "proof" that $-1 = +1$?

$$(-1)^2 = +1$$
$$\log_{10}(-1)^2 = \log_{10}(+1)$$
$$2 \log_{10}(-1) = \log_{10}(+1)$$
$$2 \log_{10}(-1) = 0 \qquad (\text{since } \log_{10} 1 = 0)$$
$$\log_{10}(-1) = 0$$
$$\log_{10}(-1) = \log_{10}(+1)$$

Thus,

$$-1 = +1$$

6. Let $x = 100$ and $y = 10$, and show by example that

$$\frac{\log_{10} x}{\log_{10} y} \neq \log_{10} \frac{x}{y}$$

3.4 TABLES

Logarithms using base 10 are called *common logarithms*. Because they are frequently used, $\log_{10} x$ is written simply as $\log x$. In other words, when no base is specified, base 10 is assumed. The irrational number e (approximately 2.718) is also used frequently for a logarithm base, especially in calculus settings, where it arises natually. Such "natural logarithms," as they are called, are written using *ln* rather than log. Thus, $\log_e x$ is written as *ln x*. Now let's consider some examples of common logs. Here are a few that can be determined by thinking in exponential form.

Count # ZEROS

log 10,000	=	4	since $10^4 = 10,000$
log 1000	=	3	since $10^3 = 1000$
log 100	=	2	since $10^2 = 100$
log 10	=	1	since $10^1 = 10$
log 1	=	0	since $10^0 = 1$
log 0.1	=	−1	since $10^{-1} = 0.1$
log 0.01	=	−2	since $10^{-2} = 0.01$

Most logarithms must be looked up in a table. Just how the entries of that table are obtained is explained in Chapter 8. The table itself is in *Page 328* the Appendix (Table I), and it gives *approximate* values of common logarithms. Better approximations are available in tables that display more places for each entry. The table contains logarithms of numbers from 1.00 to 9.99. For example, log 1.74 can be found in the table by goint down the *n* column as far as 1.7 and then over to the column labcled 4. The number there is 0.2405. Thus, log 1.74 ≐ 0.2405. The table can be read in reverse; that is, if log *n* = 0.2405, then *n* ≐ 1.74. The number 1.74 is called the *antilogarithm* of 0.2405. Use the table to verify each of the following statements.

$$\log 4.91 \doteq 0.6911$$
$$\log 7.67 \doteq 0.8848$$
$$\log 2.00 \doteq 0.3010$$
$$\text{antilog } 0.9355 \doteq 8.62$$
$$\text{antilog } 0.0128 \doteq 1.03$$
$$\text{antilog } 0.8457 \doteq 7.01$$

The logarithm of a positive number N not between 1.00 and 9.99 can also be read from this table. The number N should be written in scientific notation, that is, as a number *n* between 1.00 and 9.99 times a power of 10. The logarithm of N is then the logarithm of *n* plus the power to which 10 is raised.

$$\log N = \log n \cdot 10^c$$
$$= \log n + \log 10^c$$
$$= \log n + c \qquad \text{(since } \log_{10} 10^c = c\text{)}$$
$$= c + \log n$$

Specifically,

mantissa from table

$$\log 17.4 = \log 1.74 \cdot 10^1 \doteq 1 + 0.2405 = 1.2405$$
$$\log 174 = \log 1.74 \cdot 10^2 \doteq 2 + 0.2405 = 2.2405$$
$$\log 1740 = \log 1.74 \cdot 10^3 \doteq 3 + 0.2405 = 3.2405$$

When $\log N$ is written in the form $\log N = c + \log n$, c is called the *characteristic* and $\log n$ is called the *mantissa*. The characteristic is the power of 10. The mantissa is the number obtained from the table. In $\log 1740 \doteq 3.2405$, the characteristic is 3 and the mantissa is approximately 0.2405.

Logarithms of numbers having fewer than three digits can be found in the table after supplying enough zeros to make a three-digit number.

$$\log 2.6 = \log 2.60 \doteq 0.4150$$
$$\log 53 = \log 53.0 \doteq 1.7243$$
$$\log 8 = \log 8.00 \doteq 0.9031$$

The characteristic of $\log 0.0359$ is -2, since $\log 0.0359 = \log 3.59 \cdot 10^{-2} = -2 + \log 3.59 \doteq -2 + 0.5551$. It is usually undesirable to use such a negative characteristic, because it cannot be attached to the front of the mantissa. Compare

$$\log 529 \doteq 2 + 0.7235 \doteq 2.7235$$

But

$$\log 0.0359 \doteq -2 + 0.5551 \neq -2.5551$$

Instead

$$-2 + 0.5551 = -1.4449$$

which is an undesirable form because the mantissa has been altered. In order to preserve the mantissa, write the characteristic -2 as $8 - 10$ and insert the mantissa between the 8 and the -10, as

$$\log 0.0359 \doteq 8.5551 - 10$$

Similarly,

$$\log 0.359 \doteq 9.5551 - 10$$
$$\log 0.00359 \doteq 7.5551 - 10$$

After the *antilogarithm* of a mantissa is obtained from the table, it should be multiplied by 10^c, where c is the characteristic of the logarithm. For example, antilog $2.8267 = $ (antilog $0.8267) \cdot 10^2 \doteq 6.71 \cdot 10^2 = 671$.

Logarithms of numbers having four or more digits can be found by using the table and a method called *linear interpolation*. Consider, for example, determining the logarithm of 14.83. The logarithms of 14.8 and 14.9 are available directly from the table. The logarithm of 14.83 is between them.

mantissa from table 1.48×10^1

$$\log 14.80 \doteq 1.1703$$
$$\log 14.83 = \qquad \text{\textit{mantissa from table}} \quad 1.49 \times 10^1$$
$$\log 14.90 \doteq 1.1732$$

Log 14.83 is approximately 0.3 of the way from log 14.80 to log 14.90, that is, about 0.3 of the way from 1.1703 to 1.1732. In other words,

$$\log 14.83 \doteq \log 14.80 + 0.3 \, (\log 14.90 - \log 14.80)$$
$$\doteq 1.1703 + 0.3 \, (1.1732 \overset{.0029}{-} 1.1703)$$
$$\doteq 1.1703 + 0.3 \, (0.0029)$$
$$\doteq 1.1703 + 0.00087$$
$$\doteq 1.17117 \quad \text{or} \quad 1.1712 \quad \checkmark$$

Here is a brief explanation of what is happening in linear interpolation. The two known logarithms are approximate functional values for $f(x) = \log x$. The curve passing through them is indeed a curved line, but in small intervals it is approximately a straight line. Linear interpolation techniques approximate the curve between the two points by a straight line between the two points. The number obtained by the interpolation is from the line rather than from the actual logarithm curve. A graphic illustration of this process is shown in the drawing at the top of the next page.

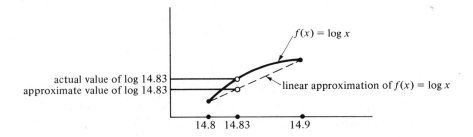

Often the *antilog* cannot be determined directly from the table; interpolation is needed. For example, the antilogarithm of 2.8930 must be found by interpolation, since the mantissa 0.8930 is not in the table. The mantissa 0.8930 is located between 0.8927 and 0.8932.

$$2.8927 \doteq \log 781.0$$
$$2.8930 = \log 781.?$$
$$2.8932 \doteq \log 782.0$$

Note that 0.8930 is about $\frac{3}{5}$, or 0.6, of the way from 0.8927 to 0.8932, because there is 0.0003 unit from 0.8927 to 0.8930 and 0.0005 unit from 0.8927 to 0.8932. The fraction 0.0003/0.0005 reduces to $\frac{3}{5}$, or 0.6. Thus, the antilog of 2.8930 is approximately 781.6. The digits of the antilog are 7816. The characteristic of 2 determines the decimal point.

EXERCISE SET D

1. Find the approximate value of each logarithm.
  *(a) log 38.6   *(b) log 491
  *(c) log 5.4   (d) log 984
   (e) log 13,000   *(f) log 0.834
   (g) log 1.5   (h) log 12.9
  *(i) log 0.0065

2. Find the approximate value of the antilogarithm of each number.
  *(a) 2.4116   *(b) 1.9335
  *(c) 3.1523   *(d) 9.7419 − 10
   (e) 1.7016   (f) 4.3997

   *Answers to starred problems are to be found at the back of the book.

*(g) 7.8129 − 10   *(h) 0.4829
(i) 8.2695 − 10

3. Find the approximate value of each logarithm. Use interpolation.
   *(a) log 489.2   *(b) log 3415
   *(c) log 29.63   (d) log 1.599
   *(e) log 2316   (f) log 59.07
   (g) log 0.01642   (h) log 5008
   *(i) log 0.0001491

4. Find the approximate value of each antilogarithm. Use interpolation.
   *(a) 1.8770   *(b) 9.7431 − 10
   (c) 4.0653   (d) 3.2934
   *(e) 2.4410   *(f) 8.5286 − 10
   (g) 9.1591 − 10   (h) 7.9055 − 10
   *(i) 0.6397

5. Simplify each expression completely.
   (a) $\log_2 32$ (b) $\log_2 2^5$
   (c) $\log_b b^7$ (d) $\log_b b^x$

6. Graph each function. State its implied domain and range.
   (a) $f(x) = \log x$ (b) $f(x) = \ln x$
   (c) $f(x) = \log(-x)$ (d) $f(x) = \log|x|$

7. The hyperbolic sine and cosine, abbreviated sinh and cosh, are defined in terms of the irrational number e.

$$\sinh x = \tfrac{1}{2}(e^x - e^{-x})$$
$$\cosh x = \tfrac{1}{2}(e^x + e^{-x})$$

Use the definitions of sinh and cosh to prove that each of the following identities is true.
   (a) $\sinh x + \cosh x = e^x$ (b) $\cosh x - \sinh x = e^{-x}$
   (c) $(\cosh x)^2 - (\sinh x)^2 = 1$

8. Solve each equation for x. (*Hint:* Use factoring.)
   *(a) $(\log x)^2 - 3\log x + 2 = 0$ [*Note:* $(\log x)^2 = (\log x)(\log x)$
                                                   $\neq 2 \log x$.]

   (b) $(\log x)^3 - (\log x)^2 - 6 \log x = 0$
   (c) $(\log x)^2 + 8 \log x + 15 = 0$

9. Solve for x: $x(3 + 7 \log_a a^x) = 0$.

*10. Solve the quadratic equation for x: $e^{2x} - e^x - 2 = 0$. Note that e is the irrational number noted earlier.

11. Solve for x: $(x + 2) \log_6 x \leq 0$. (*Hint:* Consider cases.)

12. Solve for x: $a = a_0 e^{xt} + b$.

13. Use your knowlege of logarithms to prove each statement.
    (a) $\ln e^x = x$ (*Hint:* What is $\ln e$?)
    (b) $e^{\ln x} = x$
    (c) $a^x = e^{x \ln a}$

14. Use the results of problem 13 to help simplify each expression.
    (a) $e^{3 \ln c}$
    (b) $(\ln e^t - 1)(e^{\ln t} + 1)$
    (c) $\dfrac{x^2 e^x - x e^{\ln x}}{1 - e^x}$
    (d) $\dfrac{e^{2x} + 3e^{x \ln e} + \ln e^2}{e^x + 1}$

3.5 COMPUTATION USING LOGARITHMS

Logarithms can be used to simplify some calculations. However, such use has been declining steadily since the widespread use of computers has evolved. Nevertheless, in calculus and many sciences you still need to be able to apply the properties of logarithms. This section presents some examples that show the use of logarithms for computations.

Example 5. *Compute the value of* $\log(45 \cdot 7)$.

$$\begin{aligned}
\log(45 \cdot 7) &= \log 45 + \log 7 &&\text{(by the first property of logs)} \\
&\doteq 1.6532 + 0.8451 &&\text{(from the log table)} \\
&\doteq 2.4983 \quad \checkmark
\end{aligned}$$

Example 6. *Compute the value of* $\log(295/6.3)$.

$$\begin{aligned}
\log(295/6.3) &= \log 295 - \log 6.3 &&\text{(by the second log property)} \\
&\doteq 2.4698 - 0.7993 &&\text{(from the log table)} \\
&\doteq 1.6705 \quad \checkmark
\end{aligned}$$

Example 7. *Compute the value of* $\log 34.2^6$.

$$\log 34.2^6 = 6 \cdot \log 34.2 \quad \text{(by the third property of logs)}$$
$$\doteq 6(1.5340) \quad \text{(from the log table)}$$
$$\doteq 9.2040 \quad \checkmark$$

Example 8. *Compute the value of* $\log(740 \cdot 54.3)^8$.

$$\log (740 \cdot 54.3)^8 = 8 \cdot \log (740 \cdot 54.3) \quad \text{(by the third log property)}$$
$$= 8(\log 740 + \log 54.3) \quad \text{(by the first log property)}$$
$$\doteq 8(2.8692 + 1.7348) \quad \text{(from the log table)}$$
$$\doteq 8(4.6040)$$
$$\doteq 36.8320 \quad \checkmark$$

Example 9. *Use logarithms to approximate* $\sqrt{2}$.

Let $x = \sqrt{2}$ or $2^{1/2}$. Then

$$\log x = \log 2^{1/2}$$
$$= \tfrac{1}{2} \cdot \log 2$$
$$\doteq \tfrac{1}{2}(0.3010)$$

or

$$\log x \doteq 0.1505$$
$$x \doteq \text{antilog } (0.1505)$$
$$x \doteq 1.414 \quad \checkmark \quad \text{(from the log table)}$$

Example 10. *Use logarithms to calculate* $\dfrac{(51)(4.1)^2}{10.6}$.

Let

$$x = \frac{(51)(4.1)^2}{10.6}$$

Then

$$\log x = \log \frac{(51)(4.1)^2}{10.6}$$
$$= \log (51)(4.1)^2 - \log 10.6 \quad \text{(by the second log property)}$$
$$= \log 51 + \log (4.1)^2 - \log 10.6 \quad \text{(by the first log property)}$$
$$= \log 51 + 2 \cdot \log 4.1 - \log 10.6 \quad \text{(by the third log property)}$$
$$\doteq 1.7076 + 2(0.6128) - 1.0253 \quad \text{(from the log table)}$$
$$\doteq 1.7076 + 1.2256 - 1.0253$$

Thus
$$\log x \doteq 1.9079$$
$$x \doteq \text{antilog} \ (1.9079)$$
$$x \doteq 80.9 \ \checkmark \qquad \text{(from the log table)}$$

EXERCISE SET E

1. Evaluate each expression by using logarithms.

*(a) $\sqrt{3}$ (b) $\sqrt{5}$

*(c) $\sqrt[3]{119}$ (d) $\sqrt[3]{2351}$

*(e) $(9.26)(43)$ (f) $(83.2)(1.7)(90.15)$

*(g) $\dfrac{\sqrt{341}}{3}$ (h) $\dfrac{1580}{\sqrt{312}}$

*(i) $(6.82)^2(15)$ *(j) $\dfrac{(29)(19.6)}{(1.07)(2.3)}$

*(k) $\sqrt{\dfrac{58.64}{2.65}}$ *(l) $6(2.07)^9$

*(m) $\sqrt[3]{(9.7)^2(12.2)^5}$ (n) $\dfrac{\sqrt{22.9}}{\sqrt[3]{3.016}}$

*(o) $(17)^{2/3}(15)^{4/5}$ (p) $(10)\sqrt[6]{(92.8)(6.73)}$

2. Determine which is larger, e^π or π^e. Use $e \doteq 2.72$ and $\pi \doteq 3.14$.

*3. Suppose that you have a piece of paper $\frac{1}{1000}$ inch in thickness, and that you fold it in half 50 times. (This produces 2 thicknesses with one fold, 4 with two folds, 8 with three folds, and 2^{50} with 50 folds.) How thick is the resulting wad? You may wish to use the conversion $63,360$ inches $= 1$ mile. (*Note:* The record number of folds known to have been achieved is 9.)

4. As n increases, the number $\left(1 + \dfrac{1}{n}\right)^n$

gets closer and closer to the value of the irrational number e. Determine the value of this expression for $n = 1, 2, 5, 10, 100, 500, 2000$, and $10,000$. Use logarithms whenever you believe that they will make the computation easier.

3.6 LOGARITHMIC EQUATIONS

Some equations can be solved readily by applying properties of logarithms. Here are a few examples.

*Answers to starred problems are to be found at the back of the book.

Example 11. *Solve for x:* $2^{x+1} = 7^{4-x}$.

Since

$$2^{x+1} = 7^{4-x}$$

then

$$\log 2^{x+1} = \log 7^{4-x}$$

or

$$(x + 1) \log 2 = (4 - x) \log 7 \qquad \text{(by the third property of logs)}$$

or

$$x \log 2 + 1 \log 2 = 4 \log 7 - x \log 7$$

After collecting x terms on the left and non-x terms on the right, we have

$$x \log 2 + x \log 7 = 4 \log 7 - \log 2$$

Factor out an x on the left side.

$$x(\log 2 + \log 7) = 4 \log 7 - \log 2$$

Divide both sides by $\log 2 + \log 7$, the coefficient of x.

$$x = \frac{4 \log 7 - \log 2}{\log 2 + \log 7}$$

$$\doteq \frac{4(0.8451) - 0.3010}{0.3010 + 0.8451} \qquad \text{(from the log table)}$$

$$\doteq \frac{3.0794}{1.1461} \doteq 2.68$$

So x is approximately 2.68.

Example 12. *Solve for x:* $\log (x + 1) - \log 2x = 3$.

By applying the second property of logarithms to the left side of the equation, $\log (x + 1) - \log 2x = 3$ can be written as

$$\log \frac{x + 1}{2x} = 3$$

In exponential form this is the same as

$$10^3 = \frac{x + 1}{2x}$$

or

$$1000 = \frac{x + 1}{2x}$$

which is readily solved by a series of algebraic manipulations.

$$2000x = x + 1$$
$$1999x = 1$$
$$x = \frac{1}{1999} \quad \checkmark$$

EXERCISE SET F

1. Solve each equation.

⋆(a) $3^x = 5$ (b) $5^x = 3$

⋆(c) $2^{x+3} = 7$ (d) $4^{x-3} = 5^x$

⋆(e) $2^{x+1} = 3^x$ (f) $5^{3x-2} = 2^{x+7}$

⋆(g) $2^x \cdot 3^{x-1} = 4^{3x}$ (h) $5^x \cdot 4^{x+1} = 7^x$

2. Solve each equation for x.

⋆(a) $\log x + \log (x + 3) = 1$

⋆(b) $\log x^2 - \log 5 = \log 7 + \log 2x$

(c) $\log_4 (x + 6) - \log_4 10 = \log_4 (x - 1) - \log_4 2$

3. Solve each inequality by using logarithms.

⋆(a) $3^x < 2$ (b) $7^x > \dfrac{100}{13}$

4. Solve each equation below for x in terms of the other numbers or letters.

⋆(a) $3 \cdot 10^x = 8$ ⋆(b) $ab^x = c$

(c) $(ab)^x = c$ (d) $\dfrac{ab^x}{c} = d$

3.7 CHANGE OF BASE

The logarithm of any positive number to any base can be found in terms of logarithms to the base 10. The method is demonstrated next by an example and then generalized.

Example 13. *Compute $\log_7 372$.*

Let $x = \log_7 372$. In exponential form this equation becomes

$$7^x = 372$$

⋆Answers to starred problems are to be found at the back of the book.

The logarithms of both sides of this equation are equal; that is,

$$\log_{10} 7^x = \log_{10} 372$$

or

$$x \cdot \log_{10} 7 = \log_{10} 372$$

Divide both sides by $\log_{10} 7$ to isolate x.

$$x = \frac{\log_{10} 372}{\log_{10} 7}$$

$$\doteq \frac{2.5705}{0.8451} \qquad \begin{cases} \text{from the log table (base 10,} \\ \text{of course)} \end{cases}$$

$$\doteq 3.04 \quad \checkmark$$

Compare the original $x = \log_7 372$ with the result $x = \log_{10} 372/\log_{10} 7$. In general,

$$\boxed{\log_b N = \frac{\log_{10} N}{\log_{10} b}}$$

EXERCISE SET G

1. Compute the approximate value of each term by using a base 10 logarithm table.

   ⋆(a) $\log_5 73$ (b) $\log_7 112$

   ⋆(c) $\log_6 58.4$ (d) $\log_2 940$

   ⋆(e) $\log_{10} 1800$ (f) $\log_9 6.53$

   ⋆(g) $\log_3 29$ (h) $\log_{2.5} 761$

   ⋆(i) $\log_{6.8} 10$ (j) $\log_{9.01} 100$

⋆2. Express $\log_3 549$ in terms of logarithms to the base 8 by following the pattern of the example given for base 10.

⋆3. Determine a formula for expressing $\log_b N$ in terms of logarithms of any other base B.

⋆Answers to starred problems are to be found at the back of the book.

chapter four

---

trigonometry

4.1 INTRODUCTION

Although the early Greeks developed trigonometry to measure triangles and angles (hence the name *trigonometry*, meaning "triangle measurement"), many current applications involve neither triangles nor angles. In view of this, much of the presentation within avoids triangles. This chapter introduces trigonometric functions and provides enough background material so that trigonometry can serve you as a tool (along with algebra and geometry) in your future study of calculus.

4.2 TRIGONOMETRIC POINTS

The circle with radius equal to 1 and center at the origin is called the *unit circle*. A graph of the unit circle is given at the top of the next page.

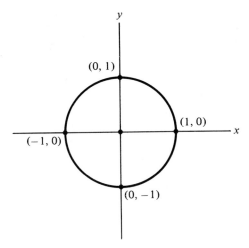

Since the equation of a circle with center at $(0, 0)$ and radius r is $x^2 + y^2 = r^2$, the equation of the unit circle is $x^2 + y^2 = 1$.†

Any point on the circle can be located by beginning at the point $(1, 0)$ and moving a counterclockwise distance t along the circle.

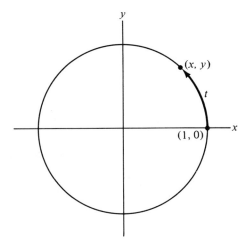

The point associated with $t = 0$ is $(1, 0)$; that is, when $t = 0$, $x = 1$ and $y = 0$. The notation $P(t)$ will be used to indicate the point, called a

---

†This statement will be proved and more will be said about circles in Chapter 6.

trigonometric point, associated with a particular *t* value. Using this notation, $P(0) = (1, 0)$.†

The circumference of a circle is $2\pi r$. For a unit circle, $r = 1$. So the distance around the unit circle is $2\pi \cdot 1 = 2\pi$‡. Thus, $P(2\pi) = (1, 0)$ just as $P(0) = (1, 0)$. Also,

$$P(0) = P(2\pi) = P(4\pi) = P(6\pi) = (1, 0)$$

In fact,

$$P(2n\pi) = (1, 0)$$

for any *whole number n.*

A negative *t* value is associated with a clockwise movement from (1, 0) along the circle.

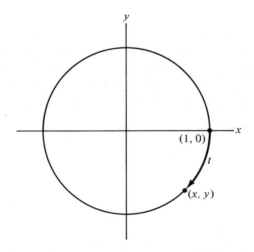

A movement of 2π units in the negative (clockwise) direction means $t = -2\pi$. The movement returns us to (1, 0) just as does a movement of 2π. Thus, $P(-2\pi) = (1, 0)$. Furthermore,

$$P(-2\pi) = P(-4\pi) = P(-6\pi) = (1, 0)$$

In fact,

$$P(2n\pi) = (1, 0)$$

for any *integer* (positive, negative, or zero) value of *n.*

---

†The range of P contains ordered pairs of real numbers.
‡The value of 2π is *approximately* 2(3.14), or 6.28.

How far must we move or measure from $(1, 0)$ to reach $(-1, 0)$? Since $(-1, 0)$ is halfway around the circle, t must be half of the circumference 2π. This means that $t = \frac{1}{2} \cdot 2\pi = \pi$. Thus,

$$P(\pi) = (-1, 0)$$

And once again you might note that moving any multiple of 2π from $P(\pi) = (-1, 0)$ will yield $(-1, 0)$. In general,

$$\boxed{P(t + 2n\pi) = P(t)}$$

for any integer n and any real number t.

To get from $(1, 0)$ to $(0, 1)$, move one-fourth of the distance around the circle. Thus, t must be $\frac{1}{4}$ of 2π, or $2\pi/4$, which reduces to $\pi/2$. So we have $P(\pi/2) = (0, 1)$. Similarly, $(0, -1)$ is three-fourths of the distance around the circle from $(1, 0)$. Since, $\frac{3}{4} \cdot 2\pi = \frac{3}{2}\pi$, $P(\frac{3}{2}\pi) = P(3\pi/2) = (0, -1)$.

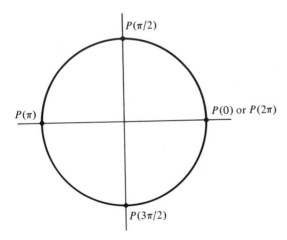

Determining the coordinates of points other than $P(0)$, $P(\pi/2)$, $P(\pi)$, and $P(3\pi/2)$ is complicated. Accordingly, they are available in tables when needed. Because they are referenced frequently we compute here $P(\pi/4)$, $P(\pi/3)$, and $P(\pi/6)$.

To compute $P(\pi/4)$, note that $P(\pi/4) = P(\frac{1}{4}\pi)$. This means that it is halfway between $P(0)$ and $P(\pi/2)$. This means that the central angle of this arc is $45°$.

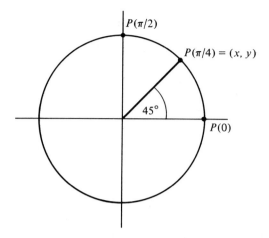

Now construct a perpendicular from $P(\pi/4)$. This produces a 45°–45°–90° triangle. The sides opposite the 45° angles are equal, since in any triangle the sides opposite two equal angles are equal.

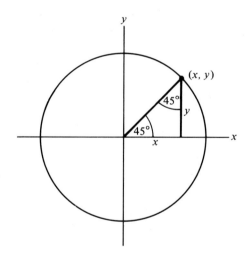

So we know that $x = y$. Thus, the coordinates (x, y) can be considered as (x, x) or as (y, y)—both the same. The relationship between x and y for all points on the circle is

$$x^2 + y^2 = 1$$

So if $x = y$, then x can be substituted for y in $x^2 + y^2 = 1$ to get

$$x^2 + x^2 = 1$$
$$2x^2 = 1$$
$$x^2 = \frac{1}{2}$$
$$x = \sqrt{\frac{1}{2}} = \frac{1}{\sqrt{2}} \quad \text{or} \quad \frac{\sqrt{2}}{2}$$

(We ignored the $\pm$ because our point is in the first quadrant and therefore both coordinates are positive.) And since $x = y$, the coordinates of $P(\pi/4)$ are both $\sqrt{2}/2$. So

$$P\left(\frac{\pi}{4}\right) = \left(\frac{\sqrt{2}}{2}, \frac{\sqrt{2}}{2}\right)$$

To compute $P(\pi/3)$, note that the arc of length $\pi/3$ is one-sixth of the circumference of the circle; that is, $\pi/3 = \frac{1}{6} \cdot 2\pi$. This means that the central angle of this arc has one-sixth the number of degrees in the circle, $\frac{1}{6}$ of 360° or 60°.

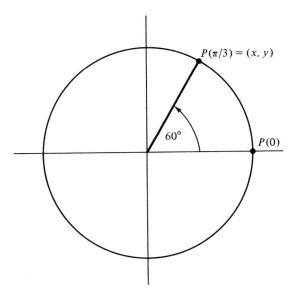

Now construct a perpendicular from $P(\pi/3)$. This produces a 30°–60°–90° triangle with sides x, y, and 1 (the radius). The drawing is shown at the top of the next page.

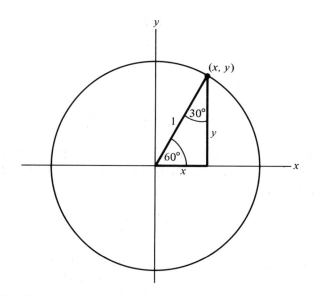

Next, we make use of the following theorem.

> ***Theorem:*** *In a 30°–60° right triangle, the side opposite the 30° angle is half the length of the hypotenuse.*

To see this, consider an equilateral triangle—one in which all sides are the same length and each of the three angles is 60°.

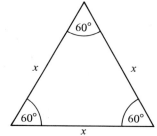

The triangle can be divided into two 30°–60°–90° triangles by drawing a perpendicular line from the top vertex to the base.

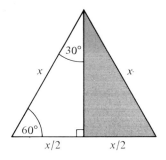

Now you can see that the base x is divided into two equal parts, each of length $x/2$. The hypotenuse is of length x. Thus, the side opposite the 30° angle in a 30°–60°–90° triangle is half the length of the hypotenuse.

Now look back at the triangle with sides x, y, and 1. Since the hypotenuse is 1 and x is opposite the 30° angle, $x = \frac{1}{2}$. To find y merely substitute $\frac{1}{2}$ for x in $x^2 + y^2 = 1$, to obtain

$$\left(\frac{1}{2}\right)^2 + y^2 = 1$$

$$\frac{1}{4} + y^2 = 1$$

$$y^2 = \frac{3}{4}$$

$$y = \sqrt{\frac{3}{4}} = \frac{\sqrt{3}}{2}$$

Thus,

$$P\left(\frac{\pi}{3}\right) = \left(\frac{1}{2}, \frac{\sqrt{3}}{2}\right)$$

The computation of $P(\pi/6)$ is similar to that of $P(\pi/3)$, because the central angle formed by an arc of length $\pi/6$ is 30°. The triangle thus formed has angles of 30°, 60°, and 90°—just as with $\pi/3$. The only difference this time is that y is opposite the 30° angle, so y is $\frac{1}{2}$. The value of x is then easily determined to be $\sqrt{3}/2$.

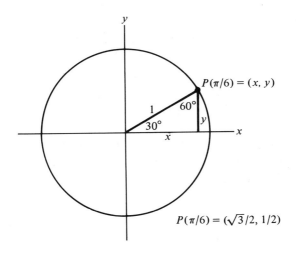

EXERCISE SET A

1. Use the results of this section and a drawing of the unit circle to determine the (x, y) coordinates of each trigonometric point.

   ★(a) $P(3\pi)$ ★(b) $P(-\pi/2)$

    (c) $P(-3\pi)$ ★(d) $P(2\pi/3)$

    (e) $P(-\pi)$ ★(f) $P(8\pi)$

    (g) $P(5\pi/2)$ (h) $P(-5\pi/2)$

   ★(i) $P(7\pi/3)$ (j) $P(-\pi/3)$

    (k) $P(-\pi/4)$ ★(l) $P(3\pi/4)$

  ★(m) $P(-3\pi/4)$ (n) $P(10\pi)$

★2. Find the length of the side opposite the 60° angle in a 30°–60°–90° triangle in which the hypotenuse is of length x. Express the answer in terms of x.

★3. Find the length of the hypotenuse of a 45°–45°–90° triangle with sides of length x. Express the answer in terms of x.

4.3 TRIGONOMETRIC FUNCTIONS

The x coordinate of the trigonometric point $P(t) = (x, y)$ is called the *cosine of t*, and it is written cos t. The y coordinate is called the *sine of t* and is written sin t. Thus,

$$x = \cos t$$
$$y = \sin t$$
$$P(t) = (x, y) = (\cos t,\ \sin t)$$

Since $P(\pi/2) = (0, 1)$ and $P(\pi/2) = (\cos \pi/2, \sin \pi/2)$, then $\cos (\pi/2) = 0$ and $\sin (\pi/2) = 1$. Similarly, $P(\pi) = (-1, 0)$ yields $\cos \pi = -1$ and $\sin \pi = 0$. Verify the following table:

★Answers to starred problems are to be found at the back of the book.

$t =$	$\dfrac{\pi}{4}$	$\dfrac{3\pi}{2}$	0	$\dfrac{\pi}{3}$
$\cos t$	$\dfrac{\sqrt{2}}{2}$	0	1	$\dfrac{1}{2}$
$\sin t$	$\dfrac{\sqrt{2}}{2}$	-1	0	$\dfrac{\sqrt{3}}{2}$

There are other trigonometric functions in addition to sine and cosine. They are defined as

$$\textbf{tangent of } t = \textbf{tan } t = \frac{y}{x} \quad \text{or} \quad \frac{\sin t}{\cos t}$$

$$\textbf{cotangent of } t = \textbf{cot } t = \frac{x}{y} \quad \text{or} \quad \frac{\cos t}{\sin t}$$

$$\textbf{secant of } t = \textbf{sec } t = \frac{1}{x} \quad \text{or} \quad \frac{1}{\cos t}$$

$$\textbf{cosecant of } t = \textbf{csc } t = \frac{1}{y} \quad \text{or} \quad \frac{1}{\sin t}$$

Note that these four trigonometric functions are defined in terms of the sine and cosine functions. The tangent and cotangent are reciprocals of each other. Cosine and secant are reciprocals. Sine and cosecant are also reciprocals. Tangent and secant are not defined for cosine (that is, x) equal to zero. Similarly, cotangent and cosecant are not defined for sine equal to zero. In each case division by zero would result.

Since $P(\pi/2) = (0, 1)$, then

$$\cos \frac{\pi}{2} = 0$$

$$\sin \frac{\pi}{2} = 1$$

$$\tan \frac{\pi}{2} = \text{undefined} \left(\frac{1}{0}\right)$$

$$\cot \frac{\pi}{2} = \frac{0}{1} = 0$$

$$\sec \frac{\pi}{2} = \text{undefined} \left(\frac{1}{0}\right)$$

$$\csc \frac{\pi}{2} = \frac{1}{1} = 1$$

Similarly, since $P(\pi/3) = (1/2, \sqrt{3}/2)$, then

$$\cos \frac{\pi}{3} = \frac{1}{2}$$

$$\sin \frac{\pi}{3} = \frac{\sqrt{3}}{2}$$

$$\tan \frac{\pi}{3} = \frac{\dfrac{\sqrt{3}}{2}}{\dfrac{1}{2}} = \sqrt{3}$$

$$\cot \frac{\pi}{3} = \frac{\dfrac{1}{2}}{\dfrac{\sqrt{3}}{2}} = \frac{1}{\sqrt{3}} \quad \text{or} \quad \frac{\sqrt{3}}{3}$$

$$\sec \frac{\pi}{3} = \frac{1}{\dfrac{1}{2}} = 2$$

$$\csc \frac{\pi}{3} = \frac{1}{\dfrac{\sqrt{3}}{2}} = \frac{2}{\sqrt{3}} \quad \text{or} \quad \frac{2\sqrt{3}}{3}$$

EXERCISE SET B

Complete the tables.

⋆1.

$t =$	0	$\dfrac{\pi}{6}$	$\dfrac{\pi}{4}$	$\dfrac{\pi}{3}$	$\dfrac{\pi}{2}$	π	$\dfrac{3\pi}{2}$
$\cos t$							
$\sin t$							
$\tan t$							
$\cot t$							
$\sec t$							
$\csc t$							

⋆Answers to starred problems are to be found at the back of the book.

2.

	$P(t)$ in quadrant:			
	I	II	III	IV
Sign of: cos t	+			
sin t	+			
tan t				
cot t				
sec t				
csc t				

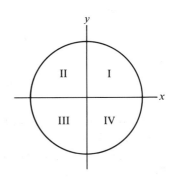

4.4 TRIGONOMETRIC IDENTITIES

If we substitute cos t for x and sin t for y in

$$x^2 + y^2 = 1$$

we get

$$(\cos t)^2 + (\sin t)^2 = 1$$

which is usually written as

$$\boxed{\sin^2 t + \cos^2 t = 1}$$

where sin^2 t means (sin t)^2 and cos^2 t means (cos t)^2.

The relationship sin^2 t + cos^2 t = 1, true for all values of t, is an example of an identity. Specifically, if a relationship (equation) is true for all values of the variable in the domain, the relationship (equation) is an *identity*.

If both sides (all terms) of the identity above are divided by sin^2 t, the result is

$$\frac{\sin^2 t}{\sin^2 t} + \frac{\cos^2 t}{\sin^2 t} = \frac{1}{\sin^2 t}$$

or

$$\boxed{1 + \cot^2 t = \csc^2 t}$$

which is an identity except for sin t = 0, in which case cot t and csc t are not defined.

If both sides of the identity $\sin^2 t + \cos^2 t = 1$ are divided by $\cos^2 t$, the result is

$$\frac{\sin^2 t}{\cos^2 t} + \frac{\cos^2 t}{\cos^2 t} = \frac{1}{\cos^2 t}$$

or

$$\boxed{\tan^2 t + 1 = \sec^2 t}$$

which is an identity except for $\cos t = 0$, in which case $\tan t$ and $\sec t$ are not defined.

The three identities derived above are often useful in other forms. For example, $\sin^2 t + \cos^2 t = 1$ can be written as $\sin^2 t = 1 - \cos^2 t$ or as $\cos^2 t = 1 - \sin^2 t$. These forms can be useful for making substitutions. Similarly, $1 + \cot^2 t = \csc^2 t$ can be written as $\csc^2 t - \cot^2 t = 1$ or as $\cot^2 t = \csc^2 t - 1$. The identity $\tan^2 t + 1 = \sec^2 t$ may be useful as $\sec^2 t - \tan^2 t = 1$ or as $\tan^2 t = \sec^2 t - 1$.

The definitions of the six trigonometric functions and the identities given above can be used to make substitutions in order to establish and prove other identities. Such proofs are usually done by changing the more complicated side of the identity step by step until it simplifies to the expression on the other side. However, sometimes it is better to change both sides of the identity to a common form. It is often helpful to reduce the tangent, cotangent, secant, and cosecant to sines and cosines. All steps of the proof must be reversible.

Example 1. *Prove the identity*

$$\frac{\sin^2 t}{1 - \sin^2 t} \stackrel{?}{=} \tan^2 t$$

(The question mark (?) indicates that we have not proved the equality; and it is removed only when the two sides have been shown to be equal.)

Change $\tan^2 t$ to $\sin^2 t / \cos^2 t$. The result is

$$\frac{\sin^2 t}{1 - \sin^2 t} \stackrel{?}{=} \frac{\sin^2 t}{\cos^2 t}$$

Since $\cos^2 t$ is the same as $1 - \sin^2 t$ (by a previous identity), replace $\cos^2 t$ by $1 - \sin^2 t$. This yields

$$\frac{\sin^2 t}{1 - \sin^2 t} = \frac{\sin^2 t}{1 - \sin^2 t} \quad \checkmark$$

We might just as well have replaced $1 - \sin^2 t$ by $\cos^2 t$ on the left side. The result would then have been

$$\frac{\sin^2 t}{\cos^2 t} = \frac{\sin^2 t}{\cos^2 t} \quad \checkmark$$

Example 2. *Prove the identity*

$$(1 - \sin t)(1 + \sin t) \overset{?}{=} \frac{1}{\sec^2 t}$$

Multiply out the $(1 - \sin t)(1 + \sin t)$ to get $1 - \sin^2 t$. The result is

$$1 - \sin^2 t \overset{?}{=} \frac{1}{\sec^2 t}$$

Replace $1 - \sin^2 t$ by $\cos^2 t$. The result is

$$\cos^2 t \overset{?}{=} \frac{1}{\sec^2 t}$$

Since $1/\sec t$ is the same as $\cos t$, replace $1/\sec^2 t$ with $\cos^2 t$. This proves the identity.

$$\cos^2 t = \cos^2 t \quad \checkmark$$

Example 3. *Prove the identity*

$$\frac{\csc t}{\cot t + \tan t} \overset{?}{=} \cos t$$

Make the following substitutions on the left side in order to change all those terms to sines and cosines:

$$\csc t = \frac{1}{\sin t}$$

$$\cot t = \frac{\cos t}{\sin t}$$

$$\tan t = \frac{\sin t}{\cos t}$$

The result is

$$\frac{\dfrac{1}{\sin t}}{\dfrac{\cos t}{\sin t} + \dfrac{\sin t}{\cos t}} \overset{?}{=} \cos t$$

Combine the fractions in the denominator of the complex fraction. The common denominator for those two fractions is $\sin t \cdot \cos t$.

$$\frac{\dfrac{1}{\sin t}}{\dfrac{\cos t}{\sin t} \cdot \dfrac{\cos t}{\cos t} + \dfrac{\sin t}{\cos t} \cdot \dfrac{\sin t}{\sin t}} \overset{?}{=} \cos t$$

$$\frac{\dfrac{1}{\sin t}}{\dfrac{\cos^2 t + \sin^2 t}{\sin t \cos t}} \overset{?}{=} \cos t$$

Substitute 1 for $\cos^2 t + \sin^2 t$.

$$\frac{\dfrac{1}{\sin t}}{\dfrac{1}{\sin t \cos t}} \overset{?}{=} \cos t$$

Carry out the division of fractions. That is, invert the denominator fraction and multiply the numerator fraction by it.

$$\frac{1}{\sin t} \cdot \frac{\sin t \cos t}{1} \overset{?}{=} \cos t$$

The $\sin t$ terms divide out, leaving simply $\cos t$ on the left. The identity is proved.

$$\cos t = \cos t \quad \checkmark$$

Example 4. *Prove the identity*

$$\tan t + \cot t \overset{?}{=} \sec^2 t \cdot \cot t$$

First, change all functions to sine and cosine.

$$\frac{\sin t}{\cos t} + \frac{\cos t}{\sin t} \overset{?}{=} \frac{1}{\cos^2 t} \cdot \frac{\cos t}{\sin t}$$

Combine the fractions on the left using the common denominator $\cos t \sin t$.

$$\frac{\sin t}{\cos t} \cdot \frac{\sin t}{\sin t} + \frac{\cos t}{\sin t} \cdot \frac{\cos t}{\cos t} \overset{?}{=} \frac{1}{\cos^2 t} \cdot \frac{\cos t}{\sin t}$$

On the left, $\sin t \cdot \sin t = \sin^2 t$ and $\cos t \cdot \cos t = \cos^2 t$. On the right, $\cos^2 t = \cos t \cdot \cos t$.

$$\frac{\sin^2 t + \cos^2 t}{\cos t \sin t} \overset{?}{=} \frac{1}{\cos t \cos t} \cdot \frac{\cos t}{\sin t}$$

On the left replace $\sin^2 t + \cos^2 t$ by 1. On the right divide out $\cos t/\cos t$.

The result is

$$\frac{1}{\cos t \sin t} = \frac{1}{\cos t \sin t} \quad \checkmark$$

Example 5. *Prove the identity*

$$\frac{1 + \sin t}{\cos t} \stackrel{?}{=} \frac{\cos t}{1 - \sin t}$$

At first it is difficult to determine just how to begin proving this identity, since all the expressions are already given in terms of sin t and cos t. Note that if $1 + \sin t$ is multiplied by $1 - \sin t$, it becomes $1 - \sin^2 t$ or $\cos^2 t$. This is a simplification, and it leads to still other simplification. Of course, if you multiply the numerator of the fraction by $1 - \sin t$, you must also multiply the denominator by $1 - \sin t$.

$$\frac{1 + \sin t}{\cos t} \cdot \frac{1 - \sin t}{1 - \sin t} \stackrel{?}{=} \frac{\cos t}{1 - \sin t}$$

$$\frac{\cos^2 t}{\cos t(1 - \sin t)} \stackrel{?}{=} \frac{\cos t}{1 - \sin t}$$

Divide out cos $t/$cos t on the left side. This completes the proof.

$$\frac{\cos t}{1 - \sin t} = \frac{\cos t}{1 - \sin t} \quad \checkmark$$

EXERCISE SET C

Prove each of the identities in problems 1–20.

1. $\tan t \cos t = \sin t$

2. $\sin t \csc t = 1$

3. $\tan x = \dfrac{\sec x}{\csc x}$

4. $\sec t = \dfrac{\tan t}{\sin t}$

5. $\cos x (\tan x + \cot x) = \csc x$

6. $\cos^2 \theta - \sin^2 \theta = 1 - 2 \sin^2 \theta$

7. $\dfrac{\tan^2 t}{\sec^2 t} = 1 - \cos^2 t$

8. $\tan \theta + \dfrac{1}{\tan \theta} = \dfrac{\csc \theta}{\cos \theta}$

9. $\csc^2 \theta(1 - \cos^2 \theta) = 1$

10. $\sec^2 x \csc^2 x = \sec^2 x + \csc^2 x$

11. $\sin \theta(\csc \theta - \sin \theta) = \cos^2 \theta$

12. $\tan t + \cot t = \csc t \sec t$

13. $(1 + \cos t)(1 - \cos t) = \tan^2 t \cos^2 t$

14. $\tan \theta \sec \theta = \dfrac{\sin \theta}{1 - \sin^2 \theta}$

15. $\dfrac{\sin x}{\csc x} + \dfrac{\cos x}{\sec x} = 1$

16. $\dfrac{1 - \cos x}{\sin x} = \dfrac{\sin x}{1 + \cos x}$

17. $\cot^2 \theta = \dfrac{\cos^2 \theta}{1 - \cos^2 \theta}$

18. $\dfrac{\sin x \cos x}{\tan x + \cot x} = \dfrac{\sin^2 x}{\sec^2 x}$

19. $\dfrac{1 + \sec t}{\tan t + \sin t} = \dfrac{1}{\sin t}$

20. $\sec x + \cot x = \dfrac{\tan x + \cos x}{\sin x}$

*21. If $\cos t = -\frac{2}{3}$ and $\sin t < 0$, find the exact values of $\sin t$, $\tan t$, $\cot t$, $\csc t$, and $\sec t$.

22. Same as problem 21, except $\sin t = \frac{1}{4}$ and $\cos t < 0$.

23. Use identities to simplify each expression as much as possible.

(a) $\tan^2 t + 1 - \sec^2 t$

(b) $4 \sin^2 t + 4 \cos^2 t$

(c) $\csc^2 t - \cot^2 t - \sin^2 t$

(d) $(1 - \cos^2 t) \cot^2 t$

(e) $\dfrac{\tan t (\tan t + \cot t)}{\cot^2 t + 1}$

24. For each problem, determine the other five trigonometric functions in terms of m. Assume $0 < t < \pi/2$.

(a) $\sin t = \dfrac{m}{3}$

(b) $\cos t = \dfrac{2}{m}$

(c) $\tan t = \dfrac{2m}{7}$

(d) $\cot t = \dfrac{\sqrt{m^2 - 9}}{3}$

(e) $\sec t = \dfrac{m}{\sqrt{m^2 - 16}}$

(f) $\csc t = \dfrac{m^2 + 1}{m}$

4.5 ANGLE MEASUREMENT

On the unit circle a central angle $\theta = 90°$ and an arc length of $t = \pi/2$ yield the same trigonometric point. If the circle has a radius of 2, then a central angle of $90°$ and an arc length of π yield the same trigonometric point. If the circle has a radius of 4, then a central angle of $90°$ and an arc length of 2π yield the same trigonometric point. Angles can be measured in units called *radians*, defined in terms of the arc length and radius of the circle.

*Answers to starred problems are to be found at the back of the book.

$$\theta = \frac{t}{r} \quad \begin{cases} \theta = \text{angle in radians} \\ t = \text{arc length} \\ r = \text{radius of circle} \end{cases}$$

From this definition, you can see that an angle of 90° is the same as an angle of $\pi/2$ radians. Consider the three cases suggested above.

$$t = \frac{\pi}{2}, \; r = 1 \qquad \theta = \frac{t}{r} = \frac{\frac{\pi}{2}}{1} = \frac{\pi}{2}$$

$$t = \pi, \; r = 2 \qquad \theta = \frac{t}{r} = \frac{\pi}{2} = \frac{\pi}{2}$$

$$t = 2\pi, \; r = 4 \qquad \theta = \frac{t}{r} = \frac{2\pi}{4} = \frac{\pi}{2}$$

If the arc length is equal to the radius r of the circle, we see that

$$\theta = \frac{r}{r} = 1 \text{ radian}$$

This means that an angle of 1 radian is the angle corresponding to an arc length equal to the radius of the circle. From $90° = \pi/2$ radians, it is easy to obtain the following results.

$$180° = \pi \text{ radians}$$

$$1° = \frac{\pi}{180} \text{ radian}$$

$$1 \text{ radian} = \frac{180}{\pi} \text{ degrees}$$

Using 3.14 as an approximate value for π, $\pi/180$ becomes approximately 0.01745. Thus,

$$1 \text{ degree} \doteq .01745 \text{ radian}$$

Similarly, $180°/\pi$ is about 57.3°. So

$$1 \text{ radian} \doteq 57.3°$$

The fractions $\pi/180°$ and $180°/\pi$ are each equal to 1, since π radians = 180°. They can be used as multipliers to change degree measure to radian measure and vice versa. The measure 225° will be changed to radians, if it is multiplied by $\pi/180°$.

$$225° = \frac{225°}{1} \cdot \frac{\pi}{180°} = \frac{5\pi}{4} \text{ radians}$$

The measure $2\pi/3$ radians will be changed to degrees, if it is multiplied by $180°/\pi$.

$$\frac{2\pi}{3} = \frac{2\pi}{3} \cdot \frac{180°}{\pi} = 120°$$

4.6 RADIUS *r*

The definitions of the trigonometric functions can be generalized for a circle of any radius r. Consider the point (x', y') on the unit circle and (x, y) on a concentric circle of radius r as shown in the figure.

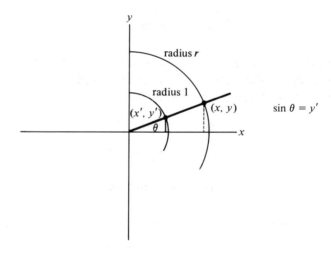

A triangle can be extracted from this setting.

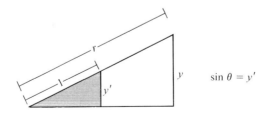

Using the proportionality of corresponding sides of similar triangles,

$$\frac{y'}{1} = \frac{y}{r}$$

or

$$y' = \frac{y}{r}$$

Now since $y' = \sin \theta$,

$$\sin \theta = y' = \frac{y}{r}$$

or

$$\boxed{\sin \theta = \frac{y}{r}}$$

Similarly,

$$\boxed{\cos \theta = \frac{x}{r}}$$

Using the definitions of the other four trigonometric functions in terms of sine and cosine, we can readily obtain

$$\boxed{\begin{aligned}\tan \theta &= \frac{y}{x} \\[4pt] \cot \theta &= \frac{x}{y} \\[4pt] \sec \theta &= \frac{r}{x} \\[4pt] \csc \theta &= \frac{r}{y}\end{aligned}}$$

4.7 TABLES

Tables of approximate values for sine, cosine, tangent, and cotangent of angles in radians and degrees are given in Tables II and III in the Appendix. How the numbers in the table are actually computed is explained in Chapter 8. In order to find the sine of an angle, for example, go down the leftmost column until you get to the desired angle.

Then move across the table until you get to the appropriate function, such as sine, for example. Here are some examples. Check each of them in Table III in the Appendix to be sure you understand how to use the table.

$$\sin 43° \doteq 0.6820$$

$$\cos 20° \doteq 0.9397$$

$$\tan 65° \doteq 2.1445$$

Here are a few examples that involve radian measure. No units are given explicitly in this setting, as is often the case with radian measure. The units are assumed to be radians unless degrees are specified. Use Table II in the Appendix.

$$\sin 1.3 \doteq 0.9636$$

$$\cos 0.2 \doteq 0.9801$$

$$\tan 1 \doteq 1.557$$

$$\sin \frac{\pi}{5} \doteq 0.5891$$

In computing sin $(\pi/5)$, change π to 3.14 and divide it by 5. This simplifies to approximately 0.63. Then you can readily look up sin 0.63 in the table.

You might argue that $\pi/5$ is not nearly 0.63; that is, $\pi/5 \doteq 3.14/5 \doteq 0.628$. But sin 0.628 is not available in the table. A more accurate value of sin $\pi/5$ can be obtained by using linear interpolation similar to that used in Chapter 3. Note that sin 0.628 is between sin 0.62 and sin 0.63.

$$\sin 0.620$$

$$\sin 0.628$$

$$\sin 0.630$$

More precisely, sin 0.628 is about $\frac{8}{10}$ of the way from sin 0.62 to sin 0.63. Now

$$\sin 0.62 \doteq 0.5810$$

and

$$\sin 0.63 \doteq 0.5891$$

So

$$\sin 0.628 \doteq \sin 0.62 + \tfrac{8}{10} (\sin 0.63 - \sin 0.62)$$

In words, sine of 0.628 is equal to the sine of 0.62 *plus* eight-tenths of the difference between the sine of 0.63 and the sine of 0.62. Thus,

$$\sin 0.628 \doteq \sin 0.62 + \tfrac{8}{10} \; (\sin 0.63 - \sin 0.62)$$
$$\doteq 0.5810 \; + \tfrac{8}{10} \; (0.5891 - 0.5810)$$
$$\doteq 0.5810 \; + \tfrac{8}{10} \; (0.0081)$$
$$\doteq 0.5810 \; + 0.0065$$
$$\doteq 0.5875 \quad \checkmark$$

4.8 TRIGONOMETRIC FUNCTIONS BEYOND QUADRANT I

Tables supply values of trigonometric functions only within the first quadrant. So trigonometric functions of t or θ for $t > \pi/2$ or $\theta > 90°$ are not directly available from tables. However, $t > \pi/2$ can be reduced to $t_1 < \pi/2$, and $\theta > 90°$ can be reduced to $\theta_1 < 90°$. Consider a point on the unit circle in the second quadrant.

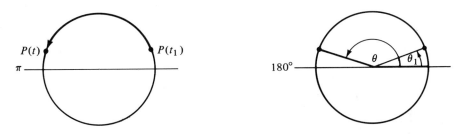

You can see that the magnitudes of the coordinates of $P(t)$ and $P(t_1)$ are the same. Specifically, $\sin t = \sin t_1$ and $\cos t = -\cos t_1$. Similarly, $\sin \theta = \sin \theta_1$ and $\cos \theta = -\cos \theta_1$. Note also that

$$t_1 = \pi - t$$
$$\theta_1 = 180° - \theta$$

Thus,

$$\left.\begin{array}{l} \sin t = \sin (\pi - t) \\ \cos t = -\cos (\pi - t) \\ \sin \theta = \sin (180° - \theta) \\ \cos \theta = -\cos (180° - \theta) \end{array}\right\} \text{quadrant II}$$

Illustrations and studies can also be made for the third and fourth quadrants. The conclusions are:

$$\left.\begin{array}{l} \sin t = -\sin (t - \pi) \\ \cos t = -\cos (t - \pi) \\ \sin \theta = -\sin (\theta - 180°) \\ \cos \theta = -\cos (\theta - 180°) \end{array}\right\} \text{quadrant III}$$

and

$$\left.\begin{array}{l} \sin t = -\sin (2\pi - t) \\ \cos t = \cos(2\pi - t) \\ \sin \theta = -\sin (360° - \theta) \\ \cos \theta = \cos(360° - \theta) \end{array}\right\} \text{quadrant IV}$$

In summary,

> For trigonometric functions of t (that is, in radians) outside quadrant I, refer t to π or 2π as above. If necessary, adjust the sign of the new form.

> For trigonometric functions of θ (degrees) outside quadrant I, refer θ to 180° or 360° as above. If necessary, adjust the sign of the new form.

The pattern of signs for sine and cosine in each quadrant is

	I	II	III	IV
sine	+	+	−	−
cosine	+	−	−	+

Signs for any other trigonometric functions are easily derived from these signs and the definition of the function.

Here are some examples showing the use of the rules.

$$\sin 150° = \sin(180° - 150°) = \sin 30° = \frac{1}{2}$$

$$\cos 135° = -\cos(180° - 135°) = -\cos 45° = -\frac{\sqrt{2}}{2}$$

$$\tan 210° = \tan(210° - 180°) = \tan 30° = \frac{\sqrt{3}}{3}$$

$$\cos 240° = -\cos(240° - 180°) = -\cos 60° = -\frac{1}{2}$$

$$\sin 315° = -\sin(360° - 315°) = -\sin 45° = -\frac{\sqrt{2}}{2}$$

$$\sin \frac{7\pi}{6} = -\sin\left(\frac{7\pi}{6} - \pi\right) = -\sin \frac{\pi}{6} = -\frac{1}{2}$$

$$\cot \frac{2\pi}{3} = -\cot\left(\pi - \frac{2\pi}{3}\right) = -\cot \frac{\pi}{3} = -\frac{\sqrt{3}}{3}$$

$$\cos \frac{5\pi}{4} = -\cos\left(\frac{5\pi}{4} - \pi\right) = -\cos \frac{\pi}{4} = -\frac{\sqrt{2}}{2}$$

$$\csc \frac{11\pi}{6} = -\csc\left(2\pi - \frac{11\pi}{6}\right) = -\csc \frac{\pi}{6} = -2$$

$$\sec \frac{7\pi}{4} = \sec\left(2\pi - \frac{7\pi}{4}\right) = \sec \frac{\pi}{4} = \sqrt{2}$$

For $\theta > 360°$, subtract as many multiples of 360° as necessary to produce a new $\theta < 360°$. Then apply the rules above as needed. Similarly, for $t > 2\pi$, subtract as many multiples of 2π as necessary to produce $t < 2\pi$. Then apply the rules above as needed. This follows from the previously shown property, $P(t + 2n\pi) = P(t)$. Here are two examples:

$$\cos 570° = \cos(570° - 360°) = \cos 210° = -\cos(210° - 180°)$$
$$= -\cos 30° = -\frac{\sqrt{3}}{2}$$

$$\sin \frac{19\pi}{4} = \sin 4\frac{3}{4}\pi = \sin\left(4\frac{3}{4}\pi - 4\pi\right) = \sin \frac{3\pi}{4} = \sin\left(\pi - \frac{3\pi}{4}\right)$$
$$= \sin \frac{\pi}{4} = \frac{\sqrt{2}}{2}$$

EXERCISE SET D

★1. Use Table III in the Appendix to determine the approximate value of each function for the given θ.

(a) $\sin 56°$ (b) $\cos 12°$

(c) $\tan 87°$ (d) $\tan 34°$

★Answers to starred problems are to be found at the back of the book.

(e) sin 30°   (f) sin 50°
(g) cos 24°   (h) cos 34°
(i) sin 389°   (j) sin 410°
(k) sin 751°   (l) cos 1100°

★2. Use Table II to determine the approximate value of each function for the given t.

(a) sin 1.2   (b) cos 0.5
(c) sin 0.35   (d) cos 1.56

(e) tan 0.23   (f) $\sin \frac{\pi}{9}$

(g) $\sin \frac{\pi}{7}$ (h) sin 3.58

(i) cos 7.59   (j) sin 10

★3. Use Table II to determine the approximate value of each function for the given t. Be sure to use interpolation.

(a) sin 1.265   (b) cos 0.674
(c) cos 0.876   (d) sin 0.447
(e) sin 1.503   (f) cos 1.009
(g) cos 0.871   (h) sin 0.777
(i) cos 0.662   (j) cos 1.222

★4. Determine the exact value of each function for the given θ. (Do not use the tables in the Appendix.)

(a) sin 210°   (b) sin 135°   (c) sin 240°
(d) cos 300°   (e) cos 120°   (f) cos 210°
(g) sin 225°   (h) sin 510°   (i) sin 330°
(j) sin 480°   (k) cos 600°   (l) sin 1050°

(m) cos 150°   (n) cos 315°   (o) $\cos \frac{8\pi}{3}$

(p) $\sin \frac{11\pi}{4}$ (q) $\cos \frac{15\pi}{4}$ (r) $\sin \frac{11\pi}{3}$

(s) $\sin \frac{11\pi}{6}$ (t) $\cos \frac{7\pi}{6}$

★5. Change each from radians to degrees or degrees to radians as appropriate.

(a) 270°   (b) 75°   (c) $\frac{\pi}{5}$

(d) $\frac{\pi}{6}$ (e) $\frac{\pi}{10}$ (f) 135°

(g) 120°   (h) 210°   (i) $\frac{3\pi}{4}$

(j) 315°   (k) 900°   (l) 225°

(m) 3π (n) $\frac{7\pi}{6}$ (o) $\frac{13\pi}{4}$

4.9 GRAPHS

Let us graph the trigonometric functions. To graph sine, use the function

$$f(t) = \sin t$$

In this way, values can be selected for t in order to get points of the form $(t, f(t))$. Follow the value of sine around the unit circle. Sine increases from 0 to 1 between 0 and $\pi/2$.

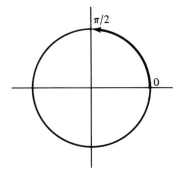

The graph thus far is

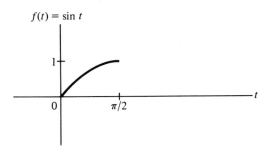

Then sine decreases from 1 to 0 between $\pi/2$ and π.

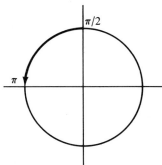

The graph now appears as

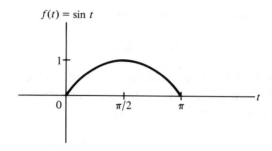

Sine decreases from 0 to -1 between π and $3\pi/2$.

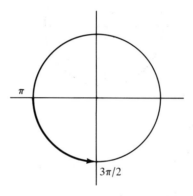

The graph now appears as

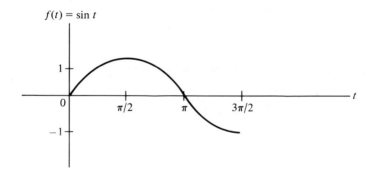

Sine increases from -1 to 0 between $3\pi/2$ and 2π.

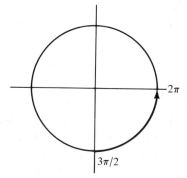

The graph now appears as

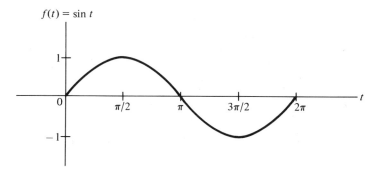

The cycle is complete. If we go around the circle again (that is, use values for t between 2π and 4π), the curve will be repeated. Similarly, if we go around the circle in the negative direction, we get additional cycles of the basic graph.

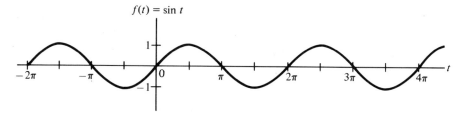

For future convenience in graphing, the function $f(t) = \sin t$ will be written as $f(x) = \sin x$ or as $y = \sin x$. In this way, elements of the domain are values for x rather than values for t, and graphing can be done in the plane traditionally labeled xy.

The function $f(x) = \sin x$ is *periodic* with period 2π; that is, it repeats its pattern of functional values every 2π. More technically,

$$\sin(x + 2\pi) = \sin x$$

and 2π is the smallest number t for which $\sin(x + t) = \sin x$.

The *amplitude*, or maximum functional value, of $y = \sin x$ is 1. The amplitude of $y = 3 \sin x$ is 3, because the values of $\sin x$ are determined and then multiplied by 3 to determine the functional values. For example, when $x = \pi/2$, $y = 3 \cdot \sin(\pi/2) = 3 \cdot 1 = 3$. The graph of $y = 3 \sin x$ between 0 and 2π is shown next.

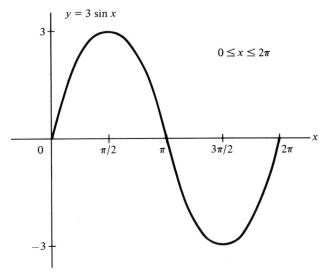

The period of $y = \sin x$ is 2π, but the period of $y = \sin 2x$ is π. The curve finishes its cycle twice as fast. Here are some sample points to show this.

x	$y = \sin x$	$y = \sin 2x$
0	$y = \sin 0 = 0$	$y = \sin 0 = 0$
$\dfrac{\pi}{4}$	$y = \sin \dfrac{\pi}{4} = \dfrac{\sqrt{2}}{2}$	$y = \sin \dfrac{\pi}{2} = 1$
$\dfrac{\pi}{2}$	$y = \sin \dfrac{\pi}{2} = 1$	$y = \sin \pi = 0$
$\dfrac{3\pi}{4}$	$y = \sin \dfrac{3\pi}{4} = \dfrac{\sqrt{2}}{2}$	$y = \sin \dfrac{3\pi}{2} = -1$
π	$y = \sin \pi = 0$	$y = \sin 2\pi = 0$

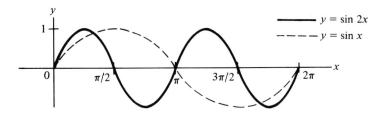

The following are additional examples.

$$y = \sin 4x \text{ has period } \frac{\pi}{2}$$

$$y = \sin \frac{1}{2}x \text{ has period } 4\pi$$

$$y = 5 \sin 2x \text{ has period } \pi, \text{ amplitude } 5$$

The more general form,

$$\boxed{y = a \sin bx}$$

describes a sine-shaped curve that has amplitude $|a|$ and period $2\pi/|b|$. The sine graph can be shifted left or right by changing its form to

$$\boxed{y = a \sin(bx + c)}$$

If $bx + c = 0$ (that is, $x = -c/b$), then y is zero. The number $-c/b$ is called the *phase shift*. It is the amount the graph of $y = a \sin bx$ must be shifted to become the graph of $y = a \sin(bx + c)$. When $-c/b < 0$ (negative), the shift is leftward. When $-c/b > 0$ (positive), the shift is to the right.

As an example, consider the graph of $y = \sin(2x + \pi)$. Its phase shift is $-\pi/2$ units. This means that the graph of $y = \sin 2x$ must be shifted $\pi/2$ units to the left to align with $y = \sin(2x + \pi)$.

$$y = \sin(2x + \pi)$$

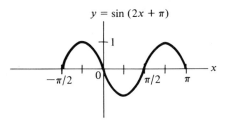

An examination of the unit circle shows that *cosine* varies from 1 to 0 between 0 and $\pi/2$, from 0 to -1 between $\pi/2$ and π, from -1 to 0 between π and $3\pi/2$, and from 0 to 1 between $3\pi/2$ and 2π. The graph of $y = \cos x$ is periodic with period 2π and amplitude 1.

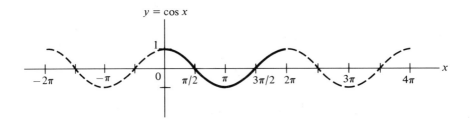

The graph of $y = \tan x$ can be sketched by observing that

$$\tan x = \frac{\sin x}{\cos x}$$

and by using the unit circle. When $x = 0$, $\sin x = 0$ and $\cos x = 1$; thus, $\tan 0 = 0/1 = 0$. Between 0 and $\pi/2$, $\sin x$ increases and $\cos x$ decreases. As a result, $\tan x$ increases. And as $\cos x$ gets close to zero near $\pi/2$, $\tan x$ gets very large. It tends toward infinity. Moreover, tangent is not defined at $\pi/2$ because cosine is zero, and division by zero is not defined. The line $x = \pi/2$ is an asymptote.

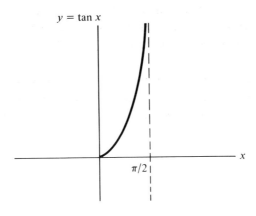

Similarly, tangent varies from $-\infty$ to 0 between $\pi/2$ and π, from 0 to ∞ between π and $3\pi/2$, and so forth.

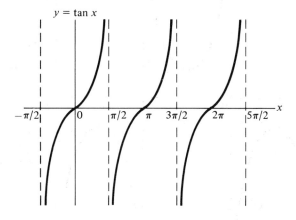

The graph of $y = \cot x$ is similar, and follows from studying $\cot x$ as $\cos x/\sin x$ or as $1/\tan x$.

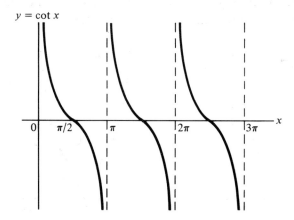

The graph of $y = \csc x$ is easily sketched once $y = \sin x$ has been drawn, since $\csc x = 1/\sin x$. Where sine is 1, cosecant is 1. As sine decreases to zero, coscant increases toward infinity. As sine increases from zero to 1, cosecant decreases from infinity to 1, and so on.

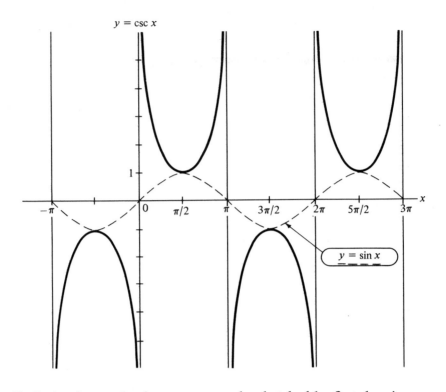

Similarly, the graph of $y = \sec x$ can be sketched by first drawing $y = \cos x$ and then using the fact that $\sec x = 1/\cos x$.

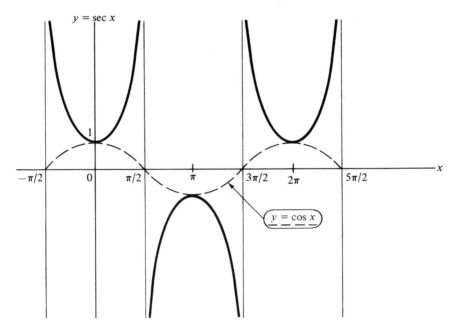

EXERCISE SET E

1. Graph each function for x values between -2π and 2π.

(a) $y = \sin x$

(b) $y = \cos x$

(c) $y = \tan x$

(d) $y = \cot x$

(e) $y = \sec x$

(f) $y = \csc x$

(g) $y = 2 \sin x$

(h) $y = \frac{1}{2} \sin x$

(i) $y = 6 \sin x$

(j) $y = \sin 4x$

(k) $y = \sin 2x$

(l) $y = \sin 3x$

(m) $y = \cos 2x$

(n) $y = 5 \cos x$

(o) $y = 3 \sin 2x$

(p) $y = 2 \sin \frac{1}{2}x$

(q) $y = 4 \cos 3x$

(r) $y = 2 \cos \frac{1}{4}x$

(s) $y = -\sin x$

(t) $y = -\cos x$

(u) $y = \sin(x + \pi)$

(v) $y = \cos\left(x + \dfrac{\pi}{2}\right)$

(w) $y = \sin(-x)$

(x) $y = -\cos(-x)$

⋆2. Determine the period and amplitude of each function in problem 1. If the amplitude is not defined, so specify.

3. Write the equations of the asymptotes of each curve.

(a) $y = \tan x$

(b) $y = \cot x$

(c) $y = \csc x$

(d) $y = \sec x$

4.10 SUMS AND DIFFERENCES

This section begins with the proof of a very important result:

$$\cos(u - v) = \cos u \cos v + \sin u \sin v$$

where u and v represent numbers of degrees or radians. This result will be used throughout the section. Note that $\cos(u - v) \neq \cos u - \cos v$. To prove the boxed formula, consider the next drawing, which shows the four trigonometric points $(\cos u, \sin u)$, $(\cos v, \sin v)$, $(\cos(u - v)$, $\sin(u - v))$, and $(1, 0)$.

⋆Answers to starred problems are to found at the back of the book.

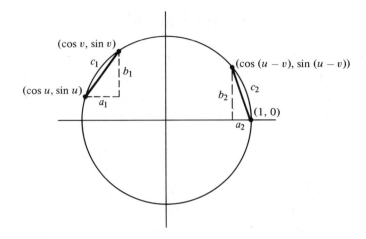

The number c_1 is the distance between (cos u, sin u) and (cos v, sin v), and c_2 is the distance between (cos($u - v$), sin($u - v$)) and (1, 0). So $c_1 = c_2$.

By the Pythagorean theorem, $c_1{}^2 = a_1{}^2 + b_1{}^2$ and $c_2{}^2 = a_2{}^2 + b_2{}^2$, where

$$|a_1| = |\cos u - \cos v|$$
$$|b_1| = |\sin u - \sin v|$$
$$|a_2| = |\cos(u - v) - 1|$$
$$|b_2| = |\sin(u - v) - 0|$$

Now

$$c_1{}^2 = (\cos u - \cos v)^2 + (\sin u - \sin v)^2$$
$$= \cos^2 u - 2 \cos u \cos v + \cos^2 v + \sin^2 u - 2 \sin u \sin v + \sin^2 v$$

Replacing $\cos^2 u + \sin^2 u$ by 1 (recall that basic identity) and $\cos^2 v + \sin^2 v$ by 1 simplifies the right side. The simplified form is

$$c_1{}^2 = 2 - 2 \cos u \cos v - 2 \sin u \sin v$$

The value of $c_2{}^2$ is equal to $c_1{}^2$, but if it is computed directly, the result has a different form. Using the Pythagorean theorem to get that form, we have

$$c_2{}^2 = [\cos(u - v) - 1]^2 + [\sin(u - v) - 0]^2$$
$$= \cos^2(u - v) - 2 \cos(u - v) + 1 + \sin^2(u - v)$$

Replacing $\cos^2(u - v) + \sin^2(u - v)$ by 1 yields

$$c_2{}^2 = 2 - 2 \cos(u - v)$$

Since $c_1{}^2 = c_2{}^2$, equate the two simplified forms obtained above. Then

$$c_1{}^2 = c_2{}^2$$

becomes

$$2 - 2 \cos u \cos v - 2 \sin u \sin v = 2 - 2 \cos(u - v)$$

Add -2 to both sides.

$$-2 \cos u \cos v - 2 \sin u \sin v = -2 \cos(u - v)$$

Interchange right and left sides.

$$-2 \cos(u - v) = -2 \cos u \cos v - 2 \sin u \sin v$$

Finally, divide both sides by -2. This will yield the desired result.

$$\cos(u - v) = \cos u \cos v + \sin u \sin v \quad \checkmark$$

Example 6. *Use the formula for* $\cos(u - v)$ *to find the exact value of* $\cos(\pi/12)$.

The number $\frac{\pi}{12}$ can be written as the difference $\frac{\pi}{3} - \frac{\pi}{4}$, to which the formula can be applied.

$$\cos(u - v) = \cos u \cos v + \sin u \sin v$$

becomes

$$\cos\left(\frac{\pi}{3} - \frac{\pi}{4}\right) = \cos \frac{\pi}{3} \cos \frac{\pi}{4} + \sin \frac{\pi}{3} \sin \frac{\pi}{4}$$

$$= \frac{1}{2} \cdot \frac{\sqrt{2}}{2} + \frac{\sqrt{3}}{2} \cdot \frac{\sqrt{2}}{2}$$

$$= \frac{\sqrt{2}}{4}(1 + \sqrt{3}) \quad \checkmark$$

If $\pi/2$ is substituted for u in the formula for $\cos(u - v)$, an interesting result is obtained.

$$\cos(u - v) = \cos u \cos v + \sin u \sin v$$

becomes

$$\cos\left(\frac{\pi}{2} - v\right) = \cos\frac{\pi}{2}\cos v + \sin\frac{\pi}{2}\sin v$$

$$= 0\cdot\cos v + 1\cdot\sin v$$

$$= \sin v$$

Thus,

$$\cos\left(\frac{\pi}{2} - v\right) = \sin v \quad \checkmark$$

If the quantity $\left(\frac{\pi}{2} - u\right)$ is used for v in $\cos\left(\frac{\pi}{2} - v\right) = \sin v$, the result is

$$\cos\left[\frac{\pi}{2} - \left(\frac{\pi}{2} - u\right)\right] = \sin\left(\frac{\pi}{2} - u\right)$$

or

$$\cos u = \sin\left(\frac{\pi}{2} - u\right) \quad \checkmark$$

Thus, we have shown so far that

$$\cos\left(\frac{\pi}{2} - v\right) = \sin v$$

$$\sin\left(\frac{\pi}{2} - v\right) = \cos v$$

or

$$\cos(90° - \theta) = \sin\theta$$

$$\sin(90° - \theta) = \cos\theta$$

Next we show that $\cos(-v) = \cos v$, by letting $u = 0$ in $\cos(u - v)$.

$$\cos(-v) = \cos(0 - v) = \cos 0 \cos v + \sin 0 \sin v$$

$$= 1\cdot\cos v + 0\cdot\sin v$$

$$= \cos v \quad \checkmark$$

To show that $\sin(-v) = -\sin v$, let $\sin(-v) = \cos\left(\frac{\pi}{2} - (-v)\right)$.

$$\sin(-v) = \cos\left(\frac{\pi}{2} - (-v)\right)$$

$$= \cos\left(\frac{\pi}{2} + v\right)$$

$$= \cos\left(v - \left(-\frac{\pi}{2}\right)\right)$$

$$= \cos v \, \cos\left(-\frac{\pi}{2}\right) + \sin v \, \sin\left(-\frac{\pi}{2}\right)$$

$$= (\cos v)(0) + (\sin v)(-1)$$

$$= -\sin v \quad \checkmark$$

Since

$$\tan v = \frac{\sin v}{\cos v}$$

then

$$\tan(-v) = \frac{\sin(-v)}{\cos(-v)} = \frac{-\sin v}{\cos v} = -\tan v \quad \checkmark$$

We have shown the following:

$$\begin{array}{rl}
\cos(-v) = & \cos v \\
\sin(-v) = & -\sin v \\
\tan(-v) = & -\tan v
\end{array}$$

A formula for $\cos(u + v)$ can be obtained from the $\cos(u - v)$ formula by writing $\cos(u + v)$ as $\cos[u - (-v)]$. Thus,

$$\cos(u + v) = \cos[u - (-v)]$$

$$= \cos u \, \cos(-v) + \sin u \, \sin(-v)$$

And since $\cos(-v) = \cos v$ and $\sin(-v) = -\sin v$,

$$\cos(u + v) = \cos u \, \cos v - \sin u \, \sin v$$

A formula for $\sin(u + v)$ can be obtained by using $\sin t = \cos\left(\frac{\pi}{2} - t\right)$, with $t = u + v$.

$$\sin(u + v) = \cos\left[\frac{\pi}{2} - (u + v)\right]$$

$$= \cos\left[\left(\frac{\pi}{2} - u\right) - v\right]$$

$$= \cos\left(\frac{\pi}{2} - u\right) \cos v + \sin\left(\frac{\pi}{2} - u\right) \sin v$$

Using previous results, we can replace $\cos\left(\frac{\pi}{2} - u\right)$ by sin u and $\sin\left(\frac{\pi}{2} - u\right)$ by cos u. The result is

$$\sin(u + v) = \sin u \cos v + \cos u \sin v$$

A formula for $\sin(u - v)$ is now easily derived from $\sin(u + v)$.

$$\begin{aligned} \sin(u - v) &= \sin[u + (-v)] \\ &= \sin u \cos(-v) + \cos u \sin(-v) \\ &= \sin u \cos v - \cos u \sin v \end{aligned}$$

since $\cos(-v) = \cos v$ and $\sin(-v) = -\sin v$. So we have shown that

$$\sin(u - v) = \sin u \cos v - \cos u \sin v$$

A formula for $\tan(u + v)$ can be derived by writing tangent as sine divided by cosine.

$$\begin{aligned} \tan(u + v) &= \frac{\sin(u + v)}{\cos(u + v)} \\ &= \frac{\sin u \cos v + \cos u \sin v}{\cos u \cos v - \sin u \sin v} \end{aligned}$$

Although this form of $\tan(u + v)$ is used occasionally, the generally more useful form is expressed in terms of tan u and tan v, and it can be obtained by dividing each term in both the numerator and denominator by cos u cos v. As a result, all terms will contain either cosine/cosine, which is one, or sine/cosine, which is tangent.

$$\begin{aligned} \tan(u + v) &= \frac{\dfrac{\sin u \cos v}{\cos u \cos v} + \dfrac{\cos u \sin v}{\cos u \cos v}}{\dfrac{\cos u \cos v}{\cos u \cos v} - \dfrac{\sin u \sin v}{\cos u \cos v}} \\[2em] &= \frac{\dfrac{\sin u}{\cos u} \cdot \dfrac{\cos v}{\cos v} + \dfrac{\cos u}{\cos u} \cdot \dfrac{\sin v}{\cos v}}{\dfrac{\cos u}{\cos u} \cdot \dfrac{\cos v}{\cos v} - \dfrac{\sin u}{\cos u} \cdot \dfrac{\sin v}{\cos v}} \\[2em] &= \frac{\tan u \cdot 1 + 1 \cdot \tan v}{1 \cdot 1 - \tan u \cdot \tan v} \end{aligned}$$

Thus

$$\tan(u + v) = \frac{\tan u + \tan v}{1 - \tan u \tan v}$$

It can also be shown that

$$\tan(u - v) = \frac{\tan u - \tan v}{1 + \tan u \tan v}$$

Sin $2t$ can be expressed in terms of $\sin t$ and $\cos t$, by letting $\sin 2t = \sin(t + t)$ and using the formula for $\sin(u + v)$ with $u = t$ and $v = t$.

$$\sin 2t = \sin(t + t)$$
$$= \sin t \cos t + \cos t \sin t$$
$$= \sin t \cos t + \sin t \cos t$$

or

$$\sin 2t = 2 \sin t \cos t$$

Similarly,

$$\cos 2t = \cos(t + t)$$
$$= \cos t \cos t - \sin t \sin t$$

or

$$\cos 2t = \cos^2 t - \sin^2 t$$

One alternative form of $\cos 2t$ can be obtained by using $1 - \sin^2 t$ for $\cos^2 t$ in $\cos 2t = \cos^2 t - \sin^2 t$.

$$\cos 2t = \cos^2 t - \sin^2 t$$
$$= 1 - \sin^2 t - \sin^2 t$$

or

$$\cos 2t = 1 - 2 \sin^2 t$$

A third form comes from letting $\sin^2 t$ be $1 - \cos^2 t$ in $\cos 2t = \cos^2 t - \sin^2 t$.

$$\cos 2t = \cos^2 t - \sin^2 t$$
$$= \cos^2 t - (1 - \cos^2 t)$$
$$= \cos^2 t - 1 + \cos^2 t$$

or

$$\cos 2t = 2 \cos^2 t - 1$$

Thus, there are three formulas for cos 2t in terms of sin t and/or cos t.

$$\cos 2t = \cos^2 t - \sin^2 t$$
$$\cos 2t = 1 - 2 \sin^2 t$$
$$\cos 2t = 2 \cos^2 t - 1$$

The formula $\cos 2t = 2 \cos^2 t - 1$ provides values for cos 2t in terms of cos t. In other words, it gives the cosine of twice an angle (or number) in terms of the cosine of the angle (or number). This relationship can be solved for cos t in terms of cos 2t. In steps:

$$\cos 2t = 2 \cos^2 t - 1$$
$$1 + \cos 2t = 2 \cos^2 t$$
$$2 \cos^2 t = 1 + \cos 2t$$
$$\cos^2 t = \frac{1 + \cos 2t}{2}$$

Finally,

$$\cos t = \pm \sqrt{\frac{1 + \cos 2t}{2}}$$

This new relationship gives the cosine of an angle in terms of the cosine of twice the angle: cos t in terms of cos 2t. If we substitute u/2 for t, the formula becomes

$$\cos \frac{u}{2} = \pm \sqrt{\frac{1 + \cos u}{2}}$$

cos u/2 is positive if P(u/2) is in quadrant I or IV. It is negative in II or III.

This form expresses the cosine of half an angle in terms of cosine of the angle. A similar formula for sine can be derived by beginning with $\cos 2t = 1 - 2 \sin^2 t$. The formula is

$$\sin \frac{u}{2} = \pm \sqrt{\frac{1 - \cos u}{2}}$$

sin u/2 is positive if P(u/2) is in quadrant I or II. It is negative in III or IV.

Using

$$\tan \frac{u}{2} = \frac{\sin \dfrac{u}{2}}{\cos \dfrac{u}{2}}$$

yields

$$\tan \frac{u}{2} = \pm\sqrt{\frac{1 - \cos u}{1 + \cos u}}$$

$\begin{cases} \tan u/2 \text{ is positive if } P(u/2) \\ \text{is in quadrant I or III. It is} \\ \text{negative in II or IV.} \end{cases}$

A formula for tan 2*t* can be obtained from tan(*u* + *v*). Begin with

$$\tan(u + v) = \frac{\tan u + \tan v}{1 - \tan u \tan v}$$

and let *u* and *v* both equal *t*. Then

$$\tan(t + t) = \frac{\tan t + \tan t}{1 - \tan t \tan t}$$

or

$$\tan 2t = \frac{2 \tan t}{1 - \tan^2 t}$$

EXERCISE SET F

*1. A value for sin 15° can be computed by applying the formula for sin (*u* − *v*) to sin 15° written as sin (45° − 30°). Carry this out, and simplify the result.

2. Use formulas for sin (*u* + *v*), cos (*u* + *v*), etc., to prove each of the following.
   (a) sin (*x* + π) = −sin *x*
   (b) cos (π + *x*) = −cos *x*
   (c) tan (*θ* + 180°) = tan *θ*
   (d) sin (*t* + 2π) = sin *t*
   (e) cos (*t* + 2π) = cos *t*
   (f) cos (*t* − π/2) = sin *t*
   (g) sin (*θ* − 90°) = −cos *θ*
   (h) sin (*θ* + 90°) = cos *θ*
   (i) cos (*θ* + 90°) = −sin *θ*
   (j) sin (*θ* + 720°) = sin *θ*

3. Show that tan (*θ* + 90°) = −cot *θ*. Use the formula that involves sines

*Answers to starred problems are to be found at the back of the book.

and cosines rather than the tangent form (Why?). Note that this is an important result and will be applied when slopes are studied in Chapter 6.

*4. (a) Derive a formula for sin $3t$ in terms of sin t. [*Hint:* Write sin $3t$ as sin $(2t + t)$ and apply the formula for sin $(u + v)$.]
   (b) Derive a formula for cos $3t$ in terms of cos t.

5. If sin $u = \frac{4}{5}$ and sin $v = \frac{5}{13}$, evaluate each of the following. Assume that u and v are in the first quadrant. Use a formula that relates sine and cosine, in order to obtain values for cos u and cos v.

*(a) sin $(u + v)$ *(b) cos $(u + v)$
 (c) sin $(u - v)$ (d) cos $(u - v)$
*(e) tan $(u + v)$ *(f) sin $2u$
 (g) sin $2v$ *(h) cos $2u$
 (i) cos $2v$ (j) tan $2u$

*6. Derive a formula for cot $(u + v)$ in terms of cot u and cot v. Then derive a formula for cot $2t$ in terms of cot t.

*7. If sin $\theta = -\frac{2}{3}$ and $270° < \theta < 360°$, determine the exact value of sin $(\theta/2)$, cos $(\theta/2)$, and tan $(\theta/2)$.

8. If sin $t = -\frac{1}{5}$ and sec $u = \frac{4}{3}$, determine the value of sin $(t + u)$ and tan $(t + u)$. Assume that t and u are both in quadrant IV.

9. Use the tests suggested in Chapter 2 to check each function for symmetry with respect to the x axis, the y axis, and the origin.

(a) $y = \sin x$ (b) $y = \cos x$
(c) $y = \tan x$ (d) $y = \cot x$
(e) $y = \sec x$ (f) $y = \csc x$
(g) $y = \sin|x|$ (h) $y = \cos|x|$
(i) $y = \tan|x|$

10. Use the formulas for sin $(u + v)$ and sin $(u - v)$ to obtain the formula

$$\sin u \cos v = \tfrac{1}{2}[\sin (u + v) + \sin (u - v)]$$

11. Use the formulas for cos $(u + v)$ and cos $(u - v)$ to obtain the results:
(a) $\cos u \cos v = \tfrac{1}{2}[\cos (u + v) + \cos (u - v)]$
(b) $\sin u \sin v = \tfrac{1}{2}[\cos (u - v) - \cos (u + v)]$

4.11 INVERSES

Can you find a value of t such that sin $t = \frac{1}{2}$? There are many solutions. In radians: $t = \pi/6, 5\pi/6, 13\pi/6, -7\pi/6$, and others. In degrees, $t = 30°$,

150°, 390°, −210°, and others. There are some settings in which we seek a unique value of t. In other words, sometimes we choose to restrict the domain of sin t so that the *inverse* we seek is unique, so that there is precisely one t. In this way, an *inverse function* is created. In the example sin $t = \frac{1}{2}$, the notation changes from

$$\sin t = \frac{1}{2}$$

to

$$\text{Sin}^{-1} \frac{1}{2} = t$$

or

$$\text{Arcsin} \frac{1}{2} = t$$

We say that t is the *Arcsine* of $\frac{1}{2}$. The notation Sin^{-1} does not denote an exponent of −1; instead it means "inverse." Note also that Sin^{-1} is capitalized.

In order to establish the inverse function for sine, we define a new function, called the *principal sine*, Sin, by restricting the domain to numbers between −π/2 and π/2 inclusive. In this way all values for sine are included only once.

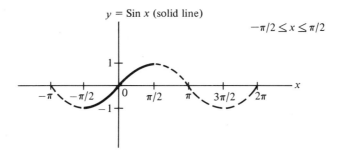

Values for Arcsin will be between −π/2 and π/2. For example,

$$\text{Arcsin} \frac{\sqrt{3}}{2} = \frac{\pi}{3}$$

$$\text{Sin}^{-1} \frac{\sqrt{2}}{2} = \frac{\pi}{4}$$

$$\text{Sin}^{-1} 1 = \frac{\pi}{2}$$

$$\text{Arcsin} 0 = 0$$

Since $\sin(-x) = -\sin x$, $\text{Arcsin}(-x) = -\text{Arcsin } x$. For example,

$$\text{Arcsin}\left(-\frac{1}{2}\right) = -\text{Arcsin } \frac{1}{2} = -\frac{\pi}{6}$$

$$\text{Arcsin}\left(-\frac{\sqrt{3}}{2}\right) = -\text{Arcsin } \frac{\sqrt{3}}{2} = -\frac{\pi}{3}$$

$$\text{Sin}^{-1}(-1) = -\text{Sin}^{-1} 1 = -\frac{\pi}{2}$$

In order to establish the inverse function for cosine, we define the *principal cosine*, Cos, by restricting the domain to numbers between 0 and π inclusive. In this way all values for cosine are included only once.

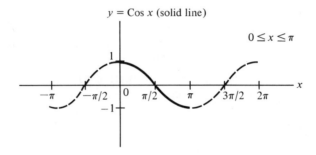

$y = \text{Cos } x$ (solid line)

$0 \le x \le \pi$

Values for Arccosine will be between 0 and π. For example,

$$\text{Arccos } \frac{\sqrt{3}}{2} = \frac{\pi}{6}$$

$$\text{Cos}^{-1} \frac{1}{2} = \frac{\pi}{3}$$

$$\text{Cos}^{-1} 1 = 0$$

$$\text{Arccos } \frac{\sqrt{2}}{2} = \frac{\pi}{4}$$

The Arccosine of a negative number is computed as π minus the Arccosine of the opposite of the number.

$$\text{Arccos}\left(-\frac{1}{2}\right) = \pi - \text{Arccos } \frac{1}{2} = \pi - \frac{\pi}{3} = \frac{2\pi}{3}$$

$$\text{Arccos}\left(-\frac{\sqrt{2}}{2}\right) = \pi - \text{Arccos } \frac{\sqrt{2}}{2} = \pi - \frac{\pi}{4} = \frac{3\pi}{4}$$

$$\text{Cos}^{-1}(-1) = \pi - \text{Cos}^{-1} 1 = \pi - 0 = \pi$$

The inverse function for tangent is established by defining the *principal tangent*, Tan, by restricting the domain to numbers between $-\pi/2$ and $\pi/2$. In this way all values for tangent are included only once.

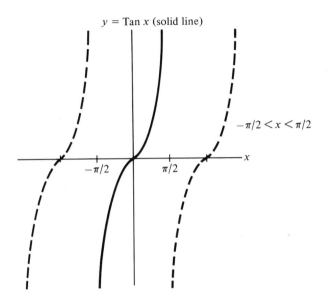

$y = \text{Tan } x$ (solid line)

$-\pi/2 < x < \pi/2$

$\text{Tan}^{-1} 1 = \pi/4$ and $\text{Arctan } \sqrt{3} = \pi/3$. Since $\tan(-x) = -\tan x$, $\text{Arctan}(-x) = -\text{Arctan } x$. So

$$\text{Arctan}(-1) = -\text{Arctan } 1 = -\frac{\pi}{4}$$

$$\text{Tan}^{-1}\left(-\frac{\sqrt{3}}{3}\right) = -\text{Tan}^{-1}\left(\frac{\sqrt{3}}{3}\right) = -\frac{\pi}{6}$$

Example 7. *Compute the value of* $\sin\left(\text{Cos}^{-1}\frac{1}{2}\right) + \tan\left(\text{Sin}^{-1}\frac{\sqrt{2}}{2}\right)$.

Since $\text{Cos}^{-1}\frac{1}{2} = \pi/3$, the expression $\sin(\text{Cos}^{-1}\frac{1}{2})$ simplifies to $\sin \pi/3$, which equals $\sqrt{3}/2$. Similarly, $\text{Sin}^{-1}\frac{\sqrt{2}}{2} = \frac{\pi}{4}$, so $\tan\left(\text{Sin}^{-1}\frac{\sqrt{2}}{2}\right)$ reduces to $\tan(\pi/4)$ or 1. Thus, the entire expression

$$\sin\left(\text{Cos}^{-1}\frac{1}{2}\right) + \tan\left(\text{Sin}^{-1}\frac{\sqrt{2}}{2}\right)$$

simplifies to

$$\frac{\sqrt{3}}{2} + 1 \quad \checkmark$$

You may prefer to write 1 as $\frac{2}{2}$ in order to combine the fractions as

$$\frac{\sqrt{3}+2}{2} \quad \text{or} \quad \frac{2+\sqrt{3}}{2} \quad \checkmark$$

Basic identities are useful in problems such as the last example, when the numbers involved are not such familiar ones.

Example 8. *Simplify* $\tan(\text{Arcsin } \frac{5}{7})$.

If $\text{Sin } t = \frac{5}{7}$, then from $\cos^2 t = 1 - \sin^2 t$, we get

$$\cos^2 t = 1 - \left(\frac{5}{7}\right)^2 = \frac{24}{49}$$

or

$$\cos t = \sqrt{\frac{24}{49}} = \frac{2\sqrt{6}}{7}$$

Note that for $\text{Sin } t = \frac{5}{7}$, t is in quadrant I. That is why we selected a positive (rather than a negative) value for the corresponding $\cos t$.

Now

$$\tan t = \frac{\sin t}{\cos t} = \frac{\dfrac{5}{7}}{\dfrac{2\sqrt{6}}{7}} = \frac{5}{2\sqrt{6}} \quad \text{or} \quad \frac{5\sqrt{6}}{12} \quad \checkmark$$

EXERCISE SET G

Simplify each expression in problems 1–20. Use Table II in the Appendix for approximations in problems 13 and 14 *only*.

★1. $\text{Arcsin } \dfrac{1}{2}$

★2. $\text{Arctan } \sqrt{3}$

★3. $\text{Arccos}\left(-\dfrac{\sqrt{3}}{2}\right)$

4. $\text{Cos}^{-1} \dfrac{\sqrt{2}}{2}$

5. $\tan\left(\text{Cos}^{-1} \dfrac{1}{2}\right)$

6. $\sin(\text{Arccos } 0)$

★7. $\sin(\text{Tan}^{-1} 1)$

★8. $\cos\left(\text{Sin}^{-1} \dfrac{1}{2}\right)$

*Answers to starred problems are to be found at the back of the book.

9. $\cos \left(\text{Tan}^{-1} (-1) \right)$

★10. $\sin \left(\text{Arcsin} \frac{1}{2} \right)$

11. $\cos \left(\text{Cos}^{-1} 1 \right)$

12. $\tan \left(\text{Sin}^{-1} 0 \right)$

★13. $\sin \left(\text{Tan}^{-1} 19.670 \right)$

★14. $\cos \left(\text{Arcsin} \ 0.2571 \right)$

★15. $\tan \left(\text{Arcsin} \frac{2}{3} \right)$

16. $\tan \left(\text{Arccos} \frac{1}{5} \right)$

★17. $\sin \left(\text{Arctan} \frac{1}{2} \right)$

18. $\csc \left(\text{Tan}^{-1} \frac{5}{12} \right)$

★19. $\cot \left(\text{Sin}^{-1} \frac{4}{5} \right)$

20. $\sec \left(\text{Tan}^{-1} \frac{1}{3} \right)$

21. Graph each inverse function by first sketching the graph of the principal trigonometric function. Then use the fact that the inverse is symmetric with respect to the line $y = x$.

(a) $f^{-1}(x) = \text{Sin}^{-1} x$

(b) $f^{-1}(x) = \text{Cos}^{-1} x$

(c) $f^{-1}(x) = \text{Tan}^{-1} x$

4.12 TRIGONOMETRIC EQUATIONS

This section consists of several examples of equations that involve trigonometric expressions. Each equation is solved.

Example 9. *Solve* $\cos x - 1 = 0$ *for* x.

If $\cos x - 1 = 0$, then $\cos x = 1$. So x is any number whose cosine is 1. Thus, $x = 0, 2\pi, 4\pi, -2\pi, -4\pi$, etc. This can be written compactly as

$$x = 2n\pi \qquad [n = \text{any integer } (0, \pm 1, \pm 2, \ldots)] \quad \checkmark$$

Example 10. *Solve* $2 \sin t + 1 = 0$ *for* t.

If $2 \sin t + 1 = 0$, then $2 \sin t = -1$, or $\sin t = -\frac{1}{2}$. So t is any number whose sine is $-\frac{1}{2}$. Thus, $t = -\pi/6, 7\pi/6$, and all other numbers determined by adding integral multiples of 2π to $-\pi/6$ and to $7\pi/6$. Thus,

$$t = -\frac{\pi}{6} + 2n\pi \qquad (n = \text{any integer}) \quad \checkmark$$

$$t = \frac{7\pi}{6} + 2n\pi \qquad (n = \text{any integer}) \quad \checkmark$$

No attempt will be made here to combine these two formulas into one compact form.

Example 11. *Solve* $\sin^2 x + 4 \sin x + 3 = 0$.

The equation is quadratic in $\sin x$. To make this more apparent, let's rewrite it as

$$(\sin x)^2 + 4(\sin x) + 3 = 0$$

The expression can be factored as

$$(\sin x + 1)(\sin x + 3) = 0$$

Thus,

$$\sin x + 1 = 0 \quad \text{or} \quad \sin x + 3 = 0$$

The equation $\sin x + 3 = 0$ leads to $\sin x = -3$, which is impossible because the sine function only has functional values between -1 and 1. On the other hand, $\sin x + 1 = 0$ leads to $\sin x = -1$ or

$$x = \frac{3\pi}{2} + 2n\pi \qquad n = \text{any integer} \quad \checkmark$$

Example 12. *Solve* $2\cos^2 x + \sin^2 x - 2 = 0$.

The identity $1 - \cos^2 x = \sin^2 x$ can be used to replace $\sin^2 x$ by $1 - \cos^2 x$. The result is an equation that is quadratic in $\cos x$. Thus,

$$2\cos^2 x + \sin^2 x - 2 = 0$$

becomes

$$2\cos^2 x + 1 - \cos^2 x - 2 = 0$$

or

$$\cos^2 x - 1 = 0$$

This leads to

$$\cos x = \pm 1$$

Now $\cos x = 1$ leads to $x = 2n\pi$, as shown in Example 9. Similarly, $\cos x = -1$ leads to $x = \pi, 3\pi, 5\pi, -\pi, -3\pi, \ldots$. In other words, $x =$ odd multiples of π. Just as $2n\pi$ was used to guarantee *even* multiples of π, $(2n + 1)\pi$ can be used to guarantee *odd* multiples of π. The solutions, then, are

$$x = 2n\pi \qquad \text{(even multiples of } \pi)$$
$$(n = \text{any integer})$$
$$x = (2n + 1)\pi \quad \text{(odd multiples of } \pi)$$

Since our solution contains all even and all odd multiples of π, it must, of course, contain *all integral multiples of* π. Thus, the two solutions above can be combined as

$$x = n\pi \quad (n = \text{any integer}) \quad \checkmark$$

EXERCISE SET H

In problems 1–18, solve each equation for all values of the variable.

★1. $\sin x = 1$

2. $2 \cos t - 1 = 0$

3. $\sin y = 0$

★4. $\cos^2 x = \dfrac{1}{4}$

★5. $\sin^2 x + 3 \sin x + 2 = 0$

6. $\sin^2 t = \dfrac{1}{2}$

★7. $\sin^2 x - \dfrac{3}{4} = 0$

8. $2 \cos^2 x - 3 \cos x + 1 = 0$

9. $2 \sin^2 x + 5 \sin x - 3 = 0$

★10. $\tan^2 u - 1 = 0$

★11. $\cos^2 x + 7 \cos x + 12 = 0$

12. $\cos^2 x + \cos x = 0$

★13. $\sin^2 x + \sin x = 0$

14. $2 \sin^2 x + \cos^2 x = 0$

15. $\cos^2 v - \sin^2 v = 0$

★16. $2 \cos^2 x - \sin^2 x + 2 \cos x = 0$

17. $\sin 2\theta = \dfrac{1}{2}$

18. $2 \cos 2\theta + 1 = 0$

19. Solve $(\sin x)(\cos x) \le 0$, for $0 \le x \le 2\pi$.

4.13 RIGHT TRIANGLES

A triangle that contains a right angle (a 90° angle) is called a *right triangle*. The next figure shows a right triangle with C being the right angle.

Side c is opposite angle C.
Side b is opposite angle B.
Side a is opposite angle A.

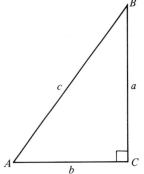

Sine, cosine, and tangent can be defined for angle θ of a right triangle in terms of the hypotenuse of the triangle and the sides opposite and adjacent to the angle θ. For any right triangle, the *hypotenuse* is the side opposite the right angle. The *adjacent side* is the side next to (adjacent to) the angle θ. The *opposite side* is the side opposite the angle θ. These terms are illustrated in the following triangle.

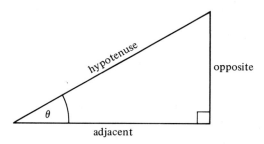

Here, then, are the definitions of sine, cosine, and tangent for a right triangle. They are expressed for the nonright angle θ. Note that these definitions are consistent with previous definitions based on circles of radius r. The positions opposite, adjacent, and hypotenuse are used instead of y, x, and r, respectively.

$$\sin \theta = \frac{\text{opposite}}{\text{hypotenuse}}$$

$$\cos \theta = \frac{\text{adjacent}}{\text{hypotenuse}}$$

$$\tan \theta = \frac{\text{opposite}}{\text{adjacent}}$$

Example 13. *Find* $\sin \theta$, $\cos \theta$, *and* $\tan \theta$ *for the given triangle.*

$$\sin \theta = \frac{\text{opposite}}{\text{hypotenuse}} = \frac{4}{5}$$

$$\cos \theta = \frac{\text{adjacent}}{\text{hypotenuse}} = \frac{3}{5}$$

$$\tan \theta = \frac{\text{opposite}}{\text{adjacent}} = \frac{4}{3}$$

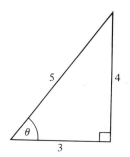

Example 14. *Find* sin ϕ, cos ϕ, *and* tan ϕ.

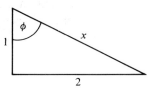

The length of the hypotenuse is not known but can be determined by using the Pythagorean theorem, $a^2 + b^2 = c^2$.

$$1^2 + 2^2 = x^2$$
$$5 = x^2$$
$$x = \sqrt{5} = \text{hypotenuse}$$

Now

$$\sin \phi = \frac{\text{opposite}}{\text{hypotenuse}} = \frac{2}{\sqrt{5}} \quad \checkmark$$

$$\cos \phi = \frac{\text{adjacent}}{\text{hypotenuse}} = \frac{1}{\sqrt{5}} \quad \checkmark$$

$$\tan \phi = \frac{\text{opposite}}{\text{adjacent}} = \frac{2}{1} = 2 \quad \checkmark$$

If you intend to approximate the values of the irrational numbers $2/\sqrt{5}$ and $1/\sqrt{5}$, rationalize the denominator of each to make the division process easier.

$$\frac{2}{\sqrt{5}} \cdot \frac{\sqrt{5}}{\sqrt{5}} = \frac{2\sqrt{5}}{5}$$

and

$$\frac{1}{\sqrt{5}} \cdot \frac{\sqrt{5}}{\sqrt{5}} = \frac{\sqrt{5}}{5}$$

Now you can use an approximation for $\sqrt{5}$ to get a rational approximation to the values of sin ϕ and cos ϕ.

Right-triangle trigonometry can be used to determine the lengths of sides of a triangle or the measure of angles of a triangle. Here are a few examples of such uses.

Example 15. *Find the length of side x in the triangle.*

The side *adjacent* to the angle of 44° is known. The side *opposite* the angle is what we seek. The trigonometric function that is defined in terms of the adjacent and opposite sides is *tangent*.

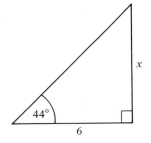

Thus,

$$\tan \theta = \frac{\text{opposite}}{\text{adjacent}}$$

becomes

$$\tan 44° = \frac{x}{6}$$

or

$$x = 6 \cdot \tan 44°$$

From the table, $\tan 44° \doteq 0.9657$. So

$$x \doteq 6(.9657)$$
$$\doteq 5.7942 \quad \checkmark$$

Example 16. *Determine the number of degrees in the angle n.*

The hypotenuse, 2, and opposite side, 1, are given. Thus,

$$\sin n = \frac{\text{opposite}}{\text{hypotenuse}} = \frac{1}{2} = 0.5000$$

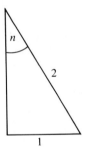

From the table, the angle whose sine is 0.5000 is 30°. Actually, this is one you should have been able to do without the table.

$$n = 30° \quad \checkmark$$

EXERCISE SET I

*1. For the right triangle shown, find the sine, cosine, and tangent of angles x and y.

2. For the right triangle shown, find the sine, cosine, and tangent of angles m and n.

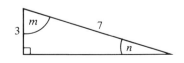

*Answers to starred problems are to be found at the back of the book.

3. For each right triangle that follows, find the approximate length of side x.

*(a)

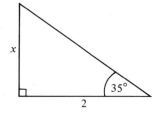

*(b)

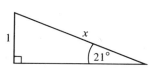

(c)

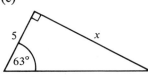

*(d)

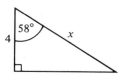

(e)

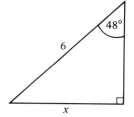

(f)

***4.** For each right triangle that follows, determine the approximate number of degrees in angle θ.

(a)

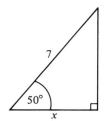

(b)

(c)

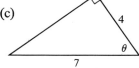

(d)

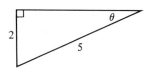

5. Verify the identity $\sin^2 \theta + \cos^2 \theta = 1$ for the given right triangle.

6. Define $\cot \theta$, $\sec \theta$, and $\csc \theta$ in terms of the hypotenuse, opposite, and adjacent sides of a right triangle.

7. Verify each identity for a right triangle with acute angle θ, adjacent side a, opposite side b, and hypotenuse c. Begin by writing each trigonometric function in terms of a, b, and c.

   (a) $\tan \theta \cos \theta = \sin \theta$
   (b) $\sin^2 \theta + \cos^2 \theta = 1$
   (c) $(1 + \cos \theta)(1 - \cos \theta) = \tan^2 \theta \cos^2 \theta$
   (d) $\tan \theta + \dfrac{1}{\tan \theta} = \dfrac{\csc \theta}{\cos \theta}$
   (e) $\tan^2 \theta + 1 = \sec^2 \theta$

8. Use right triangles to simplify each expression.

   (a) $\sin \left(\text{Tan}^{-1} \dfrac{3}{4} \right)$ (b) $\cos \left(\text{Sin}^{-1} \dfrac{5}{13} \right)$

   (c) $\tan \left(\text{Arccos} \dfrac{2}{5} \right)$ (d) $\sin \left(\text{Arccos} -\dfrac{2}{5} \right)$

   (e) $\cos \left(\text{Sin}^{-1} -\dfrac{4}{5} \right)$

9. Use the cotangent result of problem 6 and the half angle formulas in order to determine $\sin \theta$ and $\cos \theta$ corresponding to the given value of $\cot 2\theta$. A right triangle might be helpful. Assume $0 < \theta < 90$.

   (a) $\cot 2\theta = \dfrac{4}{3}$ (b) $\cot 2\theta = \dfrac{5}{12}$

   (c) $\cot 2\theta = -\dfrac{3}{4}$ (d) $\cot 2\theta = -\dfrac{1}{\sqrt{3}}$

4.14 POLAR COORDINATES

Rectangular or Cartesian coordinates have been used throughout this book. Each point is an ordered pair (x, y) in which the first coordinate, x, represents a distance left or right of the origin and the second coordinate, y, represents a distance above or below the origin. To locate a point in the plane you move over x units, then up or down y units.

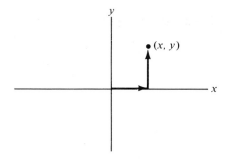

On the other hand, you *could* locate any point in the plane by moving out along the *x* axis a distance *r* and then moving in an arc an angle *θ* with respect to the *x* axis.

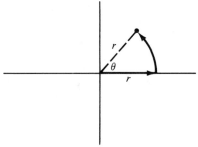

Note that $r \neq x$, as shown next.

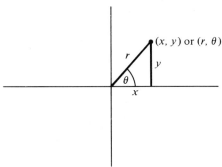

Coordinates of the form (r, θ) are called *polar coordinates*. There are several relationships between polar coordinates and rectangular coordinates. A study of the right triangle reveals them readily.

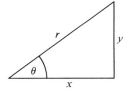

From the Pythagorean theorem,

$$x^2 + y^2 = r^2$$

or

$$r = \pm\sqrt{x^2 + y^2}$$

Also,

$$\sin \theta = \frac{y}{r} \quad \text{or} \quad y = r \sin \theta$$

$$\cos \theta = \frac{x}{r} \quad \text{or} \quad x = r \cos \theta$$

$$\tan \theta = \frac{y}{x} \quad \text{or} \quad \theta = \text{Tan}^{-1}\left(\frac{y}{x}\right)$$

The next two examples demonstrate conversion of a *point* from polar to rectangular coordinates and vice versa.

Example 17. *Change the point* $(r, \theta) = (3, \pi/6)$ *to rectangular coordinates.*

Since $x = r \cos \theta$ and $y = r \sin \theta$,

$$(x, y) = (r \cos \theta, r \sin \theta)$$

$$= \left(3 \cos \frac{\pi}{6}, 3 \sin \frac{\pi}{6}\right) \quad \text{(since } r = 3 \text{ and } \theta = \pi/6\text{)}$$

$$= \left(3 \cdot \frac{\sqrt{3}}{2}, 3 \cdot \frac{1}{2}\right)$$

$$= \left(\frac{3\sqrt{3}}{2}, \frac{3}{2}\right) \quad \checkmark$$

Example 18. *Change the point* $(x, y) = (1, \sqrt{3})$ *to polar coordinates.*

Use $r = \sqrt{x^2 + y^2}$ and $\theta = \text{Tan}^{-1}(y/x)$.

$$(r, \theta) = \left(\sqrt{x^2 + y^2}, \text{Tan}^{-1} \frac{y}{x}\right)$$

$$= (\sqrt{1 + 3}, \text{Tan}^{-1} \sqrt{3}) \quad \text{(since } x = 1 \text{ and } y = \sqrt{3}\text{)}$$

$$= \left(2, \frac{\pi}{3}\right) \quad \checkmark$$

Note that $r = +\sqrt{x^2 + y^2}$ was used rather than $r = -\sqrt{x^2 + y^2}$. For locating one specific (x, y) point, select the appropriate r, positive or negative, based on where the point is positioned. The point $(1, \sqrt{3})$ is in the first quadrant.

Here are several examples showing the techniques of changing a *relation* in rectangular coordinates (x and y) to one in polar coordinates (r and θ) and vice versa.

Example 19. *Change $y = 6x + 7$ to a relation in polar coordinates.*

Substitute $r \sin \theta$ for y and $r \cos \theta$ for x. The result is that

$$y = 6x + 7$$

becomes

$$r \sin \theta = 6r \cos \theta + 7$$
$$r \sin \theta - 6r \cos \theta = 7$$

or

$$r(\sin \theta - 6 \cos \theta) = 7$$

Finally,

$$r = \frac{7}{\sin \theta - 6 \cos \theta} \quad \checkmark$$

states the relation as r as a function of θ.

Example 20. *Change $5x^2 + 4y^2 = 15$ to a relation in polar form.*

Use $x = r \cos \theta$ and $y = r \sin \theta$. Then

$$5x^2 + 4y^2 = 15$$

becomes

$$5(r \cos \theta)^2 + 4(r \sin \theta)^2 = 15$$

or

$$5r^2 \cos^2 \theta + 4r^2 \sin^2 \theta = 15$$

Note that $5r^2 \cos^2 \theta = r^2 \cos^2 \theta + 4r^2 \cos^2 \theta$. This alternative form can be used to simplify the relation above. The new form is

$$r^2 \cos^2 \theta + 4r^2 \cos^2 \theta + 4r^2 \sin^2 \theta = 15$$

or

$$r^2 \cos^2 \theta + 4r^2 (\cos^2 \theta + \sin^2 \theta) = 15$$

You should now realize that the relation has been manipulated into this form so that substitution of 1 for $\cos^2 \theta + \sin^2 \theta$ can be made. The simplified form is then

$$r^2 \cos^2 \theta + 4r^2 = 15$$

or

$$r^2 (\cos^2 \theta + 4) = 15 \quad \checkmark$$

If desired, this relation can be solved for r^2, as

$$r^2 = \frac{15}{\cos^2 \theta + 4} \quad \checkmark$$

Example 21. *Change $r = 4$ to a relation in rectangular coordinates.*

Substitute $\pm\sqrt{x^2 + y^2}$ for r in $r = 4$ to get

$$\pm\sqrt{x^2 + y^2} = 4$$

or, upon squaring both sides,

$$x^2 + y^2 = 16 \quad \checkmark$$

Example 22. *Change $r = \dfrac{3}{1 - \cos \theta}$ to a relation in rectangular coordinates.*

Eliminate the fraction by multiplying both sides by $1 - \cos \theta$. The result is

$$r(1 - \cos \theta) = 3$$

Multiply out the left side.

$$r - r \cos \theta = 3$$

Now substitute $\pm\sqrt{x^2 + y^2}$ for r and x for $r \cos \theta$. The result is

$$\pm\sqrt{x^2 + y^2} - x = 3$$

or

$$\pm\sqrt{x^2 + y^2} = x + 3$$

After squaring both sides of this equation, we have

$$x^2 + y^2 = x^2 + 6x + 9$$

which can be written as either

$$y^2 = 6x + 9 \quad \checkmark$$

or as

$$y^2 - 6x - 9 = 0 \quad \checkmark$$

Next, relations will be graphed in polar coordinates. The polar plane is usually drawn for graphing purposes as a series of concentric circles with center at the origin and radii 1, 2, 3, 4, ... Radiating from the center are lines at angles 0°, 15°, 30°, 45°, 60°, . . . , 360° or 0, $\pi/12, \pi/6, \pi/4, \pi/3, \ldots, 2\pi$ radians.

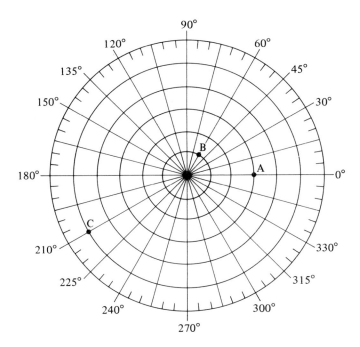

Radian measure such as 0, $\pi/6$, $\pi/4, \pi/3, \pi/2$, etc. can also be used in this setting.

Plotted above are several points:

A. (3, 0°) or (3, 0) $\qquad$ $r = 3, \theta = 0°$ or 0 radians

B. (1, 60°) or $\left(1, \dfrac{\pi}{3}\right)$ $\qquad$ $r = 1, \theta = 60°$ or $\dfrac{\pi}{3}$ radians

C. (5, 210°) or $\left(5, \dfrac{7\pi}{6}\right)$ $\qquad$ $r = 5, \theta = 210°$ or $\dfrac{7\pi}{6}$ radians

$\quad$ (−5, 30°) or $\left(-5, \dfrac{\pi}{6}\right)$ $r = -5, \theta = 30°$ or $\dfrac{\pi}{6}$ radians

Notice that the point (5, 210°) is the same as the point (−5, 30°). The first is obtained by moving 5 units to the right (positive direction) and around 210°. The second is obtained by moving 5 units to the left (negative direction) and then around 30° from that position.

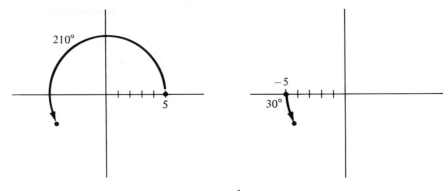

Example 23. *Sketch the graph of* $r = \dfrac{1}{\cos \theta}$.

Supply θ values in order to obtain values for r and points of the form (r, θ). A graph is shown at the top of the next page.

θ	$r = \dfrac{1}{\cos \theta}$	(r, θ)
$0°$	$\dfrac{1}{\cos 0°} = \dfrac{1}{1} = 1$	$(1, 0°)$
$30°$	$\dfrac{1}{\cos 30°} = \dfrac{1}{\sqrt{3}/2} = \dfrac{2\sqrt{3}}{3} \doteq 1.154$	$(1.154, 30°)$
$45°$	$\dfrac{1}{\cos 45°} = \dfrac{1}{\sqrt{2}/2} = \sqrt{2} \doteq 1.414$	$(1.414, 45°)$
$60°$	$\dfrac{1}{\cos 60°} = \dfrac{1}{\frac{1}{2}} = 2$	$(2, 60°)$
$90°$	$\dfrac{1}{\cos 90°} = \dfrac{1}{0} = $ undefined	—
$120°$	$\dfrac{1}{\cos 120°} = \dfrac{1}{(-\cos 60°)} = -\dfrac{1}{\cos 60°} = -2$	$(-2, 120°)$
$135°$	$\dfrac{1}{\cos 135°} = \dfrac{1}{(-\cos 45°)} = -\dfrac{1}{\cos 45°} = -1.414$	$(-1.414, 135°)$
$150°$	$\dfrac{1}{\cos 150°} = \dfrac{1}{(-\cos 30°)} = -\dfrac{1}{\cos 30°} = -1.154$	$(-1.154, 150°)$
$180°$	$\dfrac{1}{\cos 180°} = \dfrac{1}{(-1)} = -1$	$(-1, 180°)$
$210°$	$\dfrac{1}{\cos 210°} = \dfrac{1}{(-\cos 30°)} = -\dfrac{1}{\cos 30°} = -1.154$	$(-1.154, 210°)$
$225°$	$\dfrac{1}{\cos 225°} = \dfrac{1}{(-\cos 45°)} = -\dfrac{1}{\cos 45°} = -1.414$	$(-1.414, 225°)$
$240°$	$\dfrac{1}{\cos 240°} = \dfrac{1}{(-\cos 60°)} = -\dfrac{1}{\cos 60°} = -2$	$(-2, 240°)$
$270°$	$\dfrac{1}{\cos 270°} = \dfrac{1}{0} = $ undefined	—
$300°$	$\dfrac{1}{\cos 300°} = \dfrac{1}{\cos 60°} = 2$	$(2, 300°)$
$315°$	$\dfrac{1}{\cos 315°} = \dfrac{1}{\cos 45°} = 1.414$	$(1.414, 315°)$
$330°$	$\dfrac{1}{\cos 330°} = \dfrac{1}{\cos 30°} = 1.154$	$(1.154, 330°)$
$360°$	same as $0°$	$(1, 360°)$

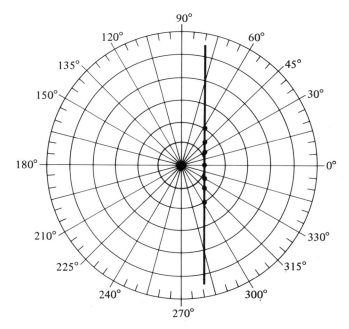

Note that several of the points
from the table are the same;
they are merely written in a
form which makes them appear
different until they are either
plotted or examined carefully.

The relation $r = 1/\cos \theta$ is a straight line. It is the line $x = 1$. This can be shown directly by writing the relation as $r \cos \theta = 1$ and then substituting x for $r \cos \theta$.

Example 24. *Sketch the graph of $r = 8 \cos \theta$.*

Again, supply θ values in order to obtain values for r. For variety the points (r, θ) are given with θ in radians this time. The graph is a circle, as shown next.

θ	$r = 8 \cos \theta$	(r, θ)
0	$r = 8 \cdot 1 = 8$	$(8, 0)$
$\dfrac{\pi}{6}$	$r = 8 \cdot \dfrac{\sqrt{3}}{2} = 4\sqrt{3}$	$\left(4\sqrt{3}, \dfrac{\pi}{6}\right)$
$\dfrac{\pi}{4}$	$r = 8 \cdot \dfrac{\sqrt{2}}{2} = 4\sqrt{2}$	$\left(4\sqrt{2}, \dfrac{\pi}{4}\right)$
$\dfrac{\pi}{3}$	$r = 8 \cdot \dfrac{1}{2} = 4$	$\left(4, \dfrac{\pi}{3}\right)$
$\dfrac{\pi}{2}$	$r = 8 \cdot 0 = 0$	$\left(0, \dfrac{\pi}{2}\right)$
$\dfrac{2\pi}{3}$	$r = 8 \cdot -\dfrac{1}{2} = -4$	$\left(-4, \dfrac{2\pi}{3}\right)$

$\dfrac{3\pi}{4}$	$r = 8 \cdot -\dfrac{\sqrt{2}}{2} = -4\sqrt{2}$	$\left(-4\sqrt{2}, \dfrac{3\pi}{4}\right)$
$\dfrac{5\pi}{6}$	$r = 8 \cdot -\dfrac{\sqrt{3}}{2} = -4\sqrt{3}$	$\left(-4\sqrt{3}, \dfrac{5\pi}{6}\right)$
π	$r = 8 \cdot -1 = -8$	$(-8, \pi)$
$\dfrac{7\pi}{6}$	$r = 8 \cdot -\dfrac{\sqrt{3}}{2} = -4\sqrt{3}$	$\left(-4\sqrt{3}, \dfrac{7\pi}{6}\right)$
$\dfrac{5\pi}{4}$	$r = 8 \cdot -\dfrac{\sqrt{2}}{2} = -4\sqrt{2}$	$\left(-4\sqrt{2}, \dfrac{5\pi}{4}\right)$
$\dfrac{4\pi}{3}$	$r = 8 \cdot -\dfrac{1}{2} = -4$	$\left(-4, \dfrac{4\pi}{3}\right)$
$\dfrac{3\pi}{2}$	$r = 8 \cdot 0 = 0$	$\left(0, \dfrac{3\pi}{2}\right)$
$\dfrac{5\pi}{3}$	$r = 8 \cdot \dfrac{1}{2} = 4$	$\left(4, \dfrac{5\pi}{3}\right)$
$\dfrac{7\pi}{4}$	$r = 8 \cdot \dfrac{\sqrt{2}}{2} = 4\sqrt{2}$	$\left(4\sqrt{2}, \dfrac{7\pi}{4}\right)$
$\dfrac{11\pi}{6}$	$r = 8 \cdot \dfrac{\sqrt{3}}{2} = 4\sqrt{3}$	$\left(4\sqrt{3}, \dfrac{11\pi}{6}\right)$
2π	$r = 8 \cdot 1 = 8$	$(8, 2\pi)$

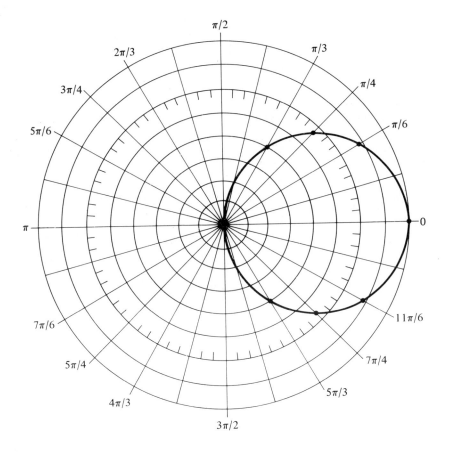

Example 25. *Sketch the graph of $r^2 = 4\cos 2\theta$.*

The figure produced when graphed is called a *lemniscate*. Shown next are a few sample points and their computation. Also given are several other points and the graph of the relation. For $\cos 2\theta < 0$, $r^2 < 0$, so there is no real r for $\cos 2\theta < 0$.

If $\theta = 0°$, then $r^2 = 4\cdot\cos(2\cdot 0°) = 4\cdot\cos 0° = 4\cdot 1 = 4; r^2 = 4$ yields $r = \pm 2$. Thus, we have the points $(2, 0°)$ and $(-2, 0°)$.

If $\theta = 30°$, then $r^2 = 4\cdot\cos(2\cdot 30°) = 4\cdot\cos 60° = 4\cdot\frac{1}{2} = 2; r^2 = 2$ yields $r = \pm\sqrt{2}$. So we have two more points—$(\sqrt{2}, 30°)$ and $(-\sqrt{2}, 30°)$.

Here are some other points: $(0, 45°)$, $(0, 135°)$, $(\sqrt{2}, 150°)$, $(-\sqrt{2}, 150°)$, $(2, 180°)$, $(-2, 180°)$, $(\sqrt{2}, 210°)$, $(-\sqrt{2}, 210°)$, $(0, 225°)$, $(0, 315°)$, $(\sqrt{2}, 330°)$, $(-\sqrt{2}, 330°)$, $(2, 360°)$, $(-2, 360°)$. Using $\theta = 15°$, $165°$, $195°$, and $345°$ will yield four other useful points. The graph is as follows:

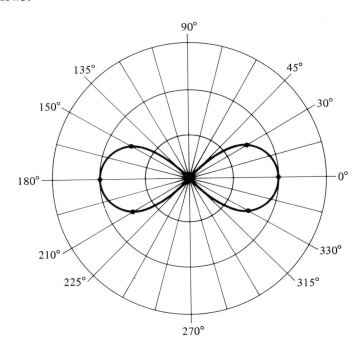

EXERCISE SET J

*★1. Change each point from rectangular coordinates to polar coordinates.
   (a) $(2, 2)$ (b) $(5, 0)$
   (c) $(\sqrt{3}, 1)$ (d) $(1, -\sqrt{3})$

*★Answers to starred problems are to be found at the back of the book.

(e) $(-5, 0)$ (f) $(0, 0)$
(g) $(-1, 1)$ (h) $(2, 1.54)$

⋆**2.** Change each point from polar coordinates to rectangular coordinates.

(a) $\left(1, \dfrac{\pi}{6}\right)$ (b) $\left(4, \dfrac{\pi}{4}\right)$

(c) $(3, 2\pi)$ (d) $(-3, 2\pi)$

(e) $\left(2, \dfrac{\pi}{2}\right)$ (f) $(0, 0)$

(g) $(1, \pi)$ (h) $(1, 1)$

3. Plot the following points and determine the nature of the curve that passes through them:

$(0, 0°)$ $\left(\dfrac{1}{2}, 30°\right)$ $\left(\dfrac{\sqrt{2}}{2}, 45°\right)$

$\left(\dfrac{\sqrt{3}}{2}, 60°\right)$ $(1, 90°)$ $\left(\dfrac{\sqrt{3}}{2}, 120°\right)$

$\left(\dfrac{\sqrt{2}}{2}, 135°\right)$ $\left(\dfrac{1}{2}, 150°\right)$ $(0, 180°)$

$\left(-\dfrac{1}{2}, 210°\right)$ $\left(-\dfrac{\sqrt{2}}{2}, 225°\right)$ $\left(-\dfrac{\sqrt{3}}{2}, 240°\right)$

$(-1, 270°)$ $\left(-\dfrac{\sqrt{3}}{2}, 300°\right)$ $\left(-\dfrac{\sqrt{2}}{2}, 315°\right)$

$\left(-\dfrac{1}{2}, 330°\right)$ $(0, 360°)$

4. Change each relation from polar coordinates to rectangular coordinates or vice versa, as appropriate.

⋆(a) $y = 2x - 3$ ⋆(b) $x^2 + y^2 = 25$
(c) $x - y = 0$ ⋆(d) $x^2 - y^2 = 16$
⋆(e) $r = 6 \sin \theta$ (f) $r = 2 \cos \theta$
(g) $3x + y - 4 = 0$ ⋆(h) $r = 5$
⋆(i) $r = \dfrac{3}{\cos \theta}$ ⋆(j) $y = x^2$

(k) $r = 1 - \cos \theta$ ⋆(l) $\theta = \dfrac{\pi}{6}$

(m) $4x^2 + 9y^2 = 36$ ⋆(n) $r = \theta$

(o) $r = \dfrac{1}{\sin \theta}$ ⋆(p) $r = 1 + \sin \theta$

5. Obtain points and graph each relation in polar coordinates. Use degrees rather than radians, unless directed otherwise.

(a) $r = 2 \sin \theta$

(b) $r = \cos \theta$

(c) $r = 3$

(d) $r = 2 \cos \theta$

(e) $r = 6 \sin \theta$

(f) $r - 4 = 0$

(g) $r = \dfrac{4}{\cos \theta}$

(h) $r \sin \theta = 16$

(i) $r \cos \theta = 2$

(j) $r = \dfrac{5}{\sin \theta}$

(k) $\theta = \dfrac{\pi}{6}$ (Use radians for this one.)

(l) $r = \dfrac{2}{1 - \cos \theta}$ (parabola)

(m) $r = 6 \sin 3\theta$ (three-leaved rose)

(n) $r = 2\theta$ (Use radians for this one.)

(o) $r = 4 \cos \theta + 2$ (limaçon)

(p) $r = 4(1 + \cos \theta)$ (cardioid)

(q) $r^2 = 5 \cos 2\theta$ (lemniscate)

(r) $r = 2 \sin \theta \cos \theta$

(s) $r = \sin^2 \theta + \cos^2 \theta$

4.15 POLAR COORDINATES AND COMPLEX NUMBERS

A complex number z can be written in the form $x + iy$. In polar coordinate form, $x = r \cos \theta$ and $y = r \sin \theta$. If $r \cos \theta$ is substituted for x and $r \sin \theta$ for y in

$$z = x + iy$$

the result is

$$z = r \cos \theta + ir \sin \theta$$

or

$$\boxed{z = r(\cos \theta + i \sin \theta)}$$

This is the *polar form* or trigonometric form of a complex number. Since for any integer k, $\cos(\theta + 2k\pi) = \cos \theta$ and $\sin(\theta + 2k\pi) = \sin \theta$,

$$z = r(\cos \theta + i \sin \theta)$$

can be considered as

$$z = r \left[\cos(\theta + 2k\pi) + i \sin(\theta + 2k\pi) \right] \qquad (k = \text{any integer})$$

Example 26. *Change $1 + i$ to polar form.*

$$z = r(\cos \theta + i \sin \theta)$$

Here

$$r = \sqrt{x^2 + y^2} = \sqrt{1^2 + 1^2} = \sqrt{2}$$

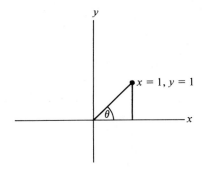

and θ satisfies $\tan \theta = y/x = 1/1$ in quadrant I, meaning $\theta = 45°$. Thus,

$$z = \sqrt{2}\,(\cos 45° + i \sin 45°) \quad \checkmark$$

or, equivalently,

$$z = \sqrt{2}\left(\cos \frac{\pi}{4} + i \sin\frac{\pi}{4}\right) \quad \checkmark$$

Example 27. *Change $4(\cos 30° + i \sin 30°)$ to rectangular form.*

$$z = 4(\cos 30° + i \sin 30°)$$
$$= 4\left(\frac{\sqrt{3}}{2} + i \cdot \frac{1}{2}\right)$$
$$= 2\sqrt{3} + 2i \quad \checkmark$$

Now consider the *product* of two complex numbers z_1 and z_2 given in polar form.

$$z_1 = r_1(\cos \theta_1 + i \sin \theta_1)$$
$$z_2 = r_2(\cos \theta_2 + i \sin \theta_2)$$
$$z_1 z_2 = [r_1(\cos \theta_1 + i \sin \theta_1)][r_2(\cos \theta_2 + i \sin \theta_2)]$$
$$= r_1 r_2(\cos\theta_1 + i \sin \theta_1)(\cos \theta_2 + i \sin \theta_2)$$
$$= r_1 r_2[(\cos \theta_1 \cos \theta_2 - \sin \theta_1 \sin \theta_2)$$
$$+ i(\sin \theta_1 \cos \theta_2 + \cos \theta_1 \sin \theta_2)]$$

Since $\cos\theta_1 \cos\theta_2 - \sin\theta_1 \sin\theta_2$ is $\cos(\theta_1 + \theta_2)$ and $\sin\theta_1 \cos\theta_2 + \cos\theta_1 \sin\theta_2$ is $\sin(\theta_1 + \theta_2)$, the following result is obtained.

$$z_1 z_2 = r_1 r_2[\cos(\theta_1 + \theta_2) + i \sin(\theta_1 + \theta_2)]$$

Example 28. *Determine the value of $z_1 \cdot z_2$, if $z_1 = 5(\cos 20° + i \sin 20°)$ and $z_2 = 8(\cos 40° + i \sin 40°)$. Write the product in polar form and in rectangular form.*

Using the formula just obtained,

$$z_1 z_2 = 5 \cdot 8[\cos (20° + 40°) + i \sin (20° + 40°)]$$
$$= 40(\cos 60° + i \sin 60°) \quad \checkmark \quad (polar\ form)$$
$$= 40\left(\frac{1}{2} + i \cdot \frac{\sqrt{3}}{2}\right)$$
$$= 20 + 20i\sqrt{3} \quad \checkmark \quad\quad\quad (rectangular\ form)$$

The *quotient* of two complex numbers z_1 and z_2 given in polar form is

$$\frac{z_1}{z_2} = \frac{r_1}{r_2}[\cos(\theta_1 - \theta_2) + i \sin(\theta_1 - \theta_2)] \quad (z_2 \neq 0)$$

Example 29. *Determine the value of z_1/z_2 for z_1 and z_2 of the previous example.*

Using the formula with

$$z_1 = 5(\cos 20° + i \sin 20°)$$

and

$$z_2 = 8(\cos 40° + i \sin 40°)$$

we obtain

$$\frac{z_1}{z_2} = \tfrac{5}{8}[\cos (20° - 40°) + i \sin (20° - 40°)]$$
$$= \tfrac{5}{8}[\cos (-20°) + i \sin (-20°)]$$
$$= \tfrac{5}{8}(\cos 20° - i \sin 20°) \quad \checkmark \quad (polar\ form)$$

Rectangular form can be obtained by using the approximations for $\cos 20°$ and $\sin 20°$ available in Table III of the Appendix.

$$\frac{z_1}{z_2} = \tfrac{5}{8}(\cos 20° - i \sin 20°)$$

$$\doteq \tfrac{5}{8}(0.9397 - i \cdot 0.3420)$$

$$\doteq 0.5873 - 0.2138i \quad \checkmark \qquad (rectangular\ form)$$

Positive integral *powers* of a complex number

$$z = r(\cos \theta + i \sin \theta)$$

are given by *De Moivre's theorem*:

$$\boxed{z^n = r^n(\cos n\theta + i \sin n\theta)}$$

Mathematical induction, explained in Chapter 8, can be used to prove that this theorem is true for all positive integers *n*. Here we shall just show that it is true for $n = 2$ and $n = 3$. The extension to higher powers should be apparent.

$$z^2 = z \cdot z = [r\,(\cos \theta + i \sin \theta)]\,[r\,(\cos \theta + i \sin \theta)]$$

$$= r \cdot r\,[\cos(\theta + \theta) + i \sin(\theta + \theta)]$$

$$= r^2\,(\cos 2\theta + i \sin 2\theta) \quad \checkmark$$

$$z^3 = z^2 z = [r^2\,(\cos 2\theta + i \sin 2\theta)]\,[r\,(\cos \theta + i \sin \theta)]$$

$$= r^2 r\,[\cos(2\theta + \theta) + i \sin (2\theta + \theta)]$$

$$= r^3\,(\cos 3\theta + i \sin 3\theta) \quad \checkmark$$

Example 30. *Use De Moivre's theorem to determine the value of* $(1 + i)^{12}$.

First, change $(1 + i)^{12}$ to polar form.

$$1 + i = r(\cos \theta + i \sin \theta)$$

$$= \sqrt{2}\left(\cos \frac{\pi}{4} + i \sin \frac{\pi}{4}\right) \qquad \begin{cases}\text{This is the result of} \\ \text{Example 26.}\end{cases}$$

so

$$(1 + i)^{12} = \left[\sqrt{2}\left(\cos \frac{\pi}{4} + i \sin \frac{\pi}{4}\right)\right]^{12}$$

$$= (\sqrt{2})^{12}\left[\cos\left(12 \cdot \frac{\pi}{4}\right) + i \sin\left(12 \cdot \frac{\pi}{4}\right)\right]$$

$$= 2^6(\cos 3\pi + i \sin 3\pi)$$

$$= 64(\cos 3\pi + i \sin 3\pi) \quad \checkmark \qquad (polar\ form)$$
$$= 64(-1 + i \cdot 0)$$
$$= -64 \quad \checkmark \qquad\qquad (rectangular\ form)$$

De Moivre's theorem can be used to prove the following result, which is used to find the nth roots of a complex number. If z is a complex number, then

$$\sqrt[n]{z} = z^{1/n} = r^{1/n}\left(\cos\frac{\theta + k \cdot 360°}{n} + i \sin\frac{\theta + k \cdot 360°}{n}\right)$$

where $k = 0, 1, 2, \ldots, n - 1$ and $r^{1/n}$ is the principal nth root of r.
The numbers

$$r^{1/n}\left(\cos\frac{\theta + k \cdot 360°}{n} + i \sin\frac{\theta + k \cdot 360°}{n}\right)$$

are indeed nth roots of z, if they yield z when raised to the nth power. The steps below verify this. By De Moivre's theorem,

$$\left[r^{1/n}\left(\cos\frac{\theta + k \cdot 360°}{n} + i \sin\frac{\theta + k \cdot 360°}{n}\right)\right]^n$$

is the same as

$$(r^{1/n})^n\left(\cos n\frac{\theta + k \cdot 360°}{n} + i \sin n\frac{\theta + k \cdot 360°}{n}\right)$$

which simplifies to z, as

$$r[\cos(\theta + k \cdot 360°) + i \sin(\theta + k \cdot 360°)]$$

and then

$$r(\cos \theta + i \sin \theta)$$

which is z in polar form.

Example 31. *Find the five fifth roots of* 1. *Leave them in polar form.*

$$z = 1 = 1 + 0i \qquad \begin{cases} r = \sqrt{1^2 + 0^2} = 1 \\ \theta = 0° \quad (\text{from } \tan \theta = y/x = 0/1 = 0) \end{cases}$$

Here $n = 5$, since there are five fifth roots of z. We shall call the roots $z_0, z_1,$

z_2, z_3, and z_4 to correspond to the k values 0 through 4. Also, the principal fifth root of 1 is 1.

$$z_k = r^{1/5}\left(\cos\frac{\theta + k \cdot 360°}{5} + i\sin\frac{\theta + k \cdot 360°}{5}\right) \qquad (k = 0, 1, 2, 3, 4)$$

so

$$z_0 = 1\left(\cos\frac{0° + 0 \cdot 360°}{5} + i\sin\frac{0° + 0 \cdot 360°}{5}\right) = 1(\cos 0° + i\sin 0°) \quad \checkmark$$

$$z_1 = 1\left(\cos\frac{0° + 1 \cdot 360°}{5} + i\sin\frac{0° + 1 \cdot 360°}{5}\right) = 1(\cos 72° + i\sin 72°) \quad \checkmark$$

$$z_2 = 1\left(\cos\frac{0° + 2 \cdot 360°}{5} + i\sin\frac{0° + 2 \cdot 360°}{5}\right) = 1(\cos 144° + i\sin 144°) \quad \checkmark$$

$$z_3 = 1\left(\cos\frac{0° + 3 \cdot 360°}{5} + i\sin\frac{0° + 3 \cdot 360°}{5}\right) = 1(\cos 216° + i\sin 216°) \quad \checkmark$$

$$z_4 = 1\left(\cos\frac{0° + 4 \cdot 360°}{5} + i\sin\frac{0° + 4 \cdot 360°}{5}\right) = 1(\cos 288° + i\sin 288°) \quad \checkmark$$

EXERCISE SET K

1. Change each number to polar form. Use Table III in the Appendix for approximations only when needed.

   *(a) $\sqrt{3} + i$ *(b) 1

   *(c) i (d) -1

   (e) $-i$ (f) $1 - i$

   *(g) $1 + i\sqrt{3}$ *(h) $2 + 2i$

   (i) $3 + 4i$ *(j) $3 - 2i$

2. Change each number from polar to rectangular form.

   *(a) $3(\cos 90° + i\sin 90°)$ (b) $10(\cos 60° + i\sin 60°)$

   *(c) $2\left(\cos\frac{\pi}{4} + i\sin\frac{\pi}{4}\right)$ (d) $6\left(\cos\frac{3\pi}{4} + i\sin\frac{3\pi}{4}\right)$

   *(e) $1\left(\cos\frac{7\pi}{4} + i\sin\frac{7\pi}{4}\right)$ *(f) $3(\cos 180° + i\sin 180°)$

   (g) $7(\cos 270° + i\sin 270°)$ *(h) $2(\cos 225° + i\sin 225°)$

   (i) $4(\cos 150° + i\sin 150°)$ (j) $7(\cos 570° + i\sin 570°)$

3. Compute each product z_1z_2 and leave it in polar form.

   *(a) $z_1 = 3(\cos 45° + i\sin 45°)$, $z_2 = 1(\cos 15° + i\sin 15°)$

*Answers to starred problems are to be found at the back of the book.

(b) $z_1 = 6(\cos 60° + i \sin 60°)$, $z_2 = 3(\cos 2° + i \sin 2°)$

★(c) $z_1 = 5\left(\cos \dfrac{\pi}{2} + i \sin \dfrac{\pi}{2}\right)$, $z_2 = 7\left(\cos \dfrac{\pi}{4} + i \sin \dfrac{\pi}{4}\right)$

(d) $z_1 = 9\left(\cos \dfrac{7\pi}{4} + i \sin \dfrac{7\pi}{4}\right)$, $z_2 = 4\left(\cos \dfrac{\pi}{4} + i \sin \dfrac{\pi}{4}\right)$

★(e) $z_1 = 1\left(\cos \dfrac{2\pi}{3} + i \sin \dfrac{2\pi}{3}\right)$, $z_2 = 2\left(\cos \dfrac{3\pi}{4} + i \sin \dfrac{3\pi}{4}\right)$

★**4.** Compute each quotient z_1/z_2 using the numbers z_1 and z_2 from problem 3. Leave the quotient in polar form.

5. Use De Moivre's theorem to compute each power.
  ★(a) $[3(\cos 30° + i \sin 30°)]^4$ (b) $[2(\cos 60° + i \sin 60°)]^5$
  ★(c) $[7(\cos 45° + i \sin 45°)]^4$ ★(d) $(1 + i\sqrt{3})^5$
  (e) $(\sqrt{3} + i)^4$ ★(f) $(i - 1)^8$
  (g) $(1 + i)^{20}$

6. Find all n nth roots for each part. Reduce $a - e$ to $x + iy$ form.
  ★(a) three cube roots of 1   (b) four fourth roots of 1
  ★(c) three cube roots of 8   ★(d) three cube roots of $-i$
  (e) four fourth roots of -1 (f) four fourth roots of $-16i$
  ★(g) four fourth roots of $1 + i$ (h) five fifth roots of $1 - i$

4.16 LAWS OF SINES AND COSINES

The following formula, called the *Law of Sines*, presents a relationship among angles and sides of *any* triangle. The triangle need not be a right triangle. If A and B are any two angles of a triangle and a and b are the sides opposite these angles, then

$$\boxed{\dfrac{\sin A}{a} = \dfrac{\sin B}{b}} \quad \text{or} \quad \boxed{\dfrac{a}{\sin A} = \dfrac{b}{\sin B}}$$

Example 32. *Find the length of the side, x, opposite the $23°$ angle.*

From the Law of Sines,

$$\dfrac{x}{\sin 23°} = \dfrac{15}{\sin 118°}$$

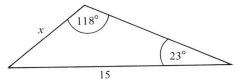

or

$$x = \frac{15\,(\sin 23°)}{\sin 118°}$$

$$\doteq \frac{15(0.3907)}{(0.8829)} \qquad \begin{cases} \sin 118° = \sin(180° - 118°) \\ \qquad = \sin 62° = 0.8829 \end{cases}$$

$$\doteq 6.6 \quad \checkmark$$

Example 33. *Determine the size of the angle opposite the side of length 12.*

$$\frac{\sin x}{12} = \frac{\sin 30°}{7}$$

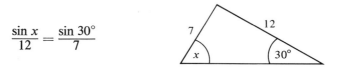

$$\sin x = \frac{12(\sin 30°)}{7} = \frac{12(0.5000)}{7} \doteq 0.8571$$

which means

$$x \doteq 59° \quad \checkmark$$

The *Law of Cosines* offers another relationship among angles and sides of a triangle. Specifically, it gives a relationship between the three sides and one angle.

$$a^2 = b^2 + c^2 - 2bc \cos A$$

where A is the angle opposite side a

The law is a generalization of the Pythagorean theorem. If A is a right angle, then $\cos A = 0$, and $a^2 = b^2 + c^2$, with a being the hypotenuse.

Derivations of both the Law of Sines and Law of Cosines are available in many trigonometry texts.

Example 34. *Find the length of the missing side.*

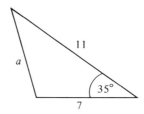

Call the missing side a. Then by the Law of Cosines,

$$a^2 = b^2 + c^2 - 2bc \cos A$$

or

$$a^2 = 11^2 + 7^2 - 2 \cdot 11 \cdot 7 \cdot \cos 35°$$

$$\doteq 121 + 49 - 154(0.8192)$$

$$\doteq 44$$

$\begin{cases}\text{We are calling } b \text{ 11 and} \\ c \text{ 7, but those two could} \\ \text{have been interchanged.}\end{cases}$

If $a^2 \doteq 44$, then $a \doteq \sqrt{44}$, or about 6.63. ✓

Example 35. *Find the value of θ in the triangle.*

Call the angle θ by the name A so that it fits the formula of the Law of Cosines. Then a is 4, and the formula can be used as

$$a^2 = b^2 + c^2 - 2bc \cos A$$

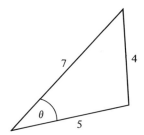

or

$$4^2 = 7^2 + 5^2 - 2 \cdot 7 \cdot 5 \cdot \cos A$$

$$16 = 49 + 25 - 70 \cdot \cos A$$

which yields

$$\cos A = \frac{-58}{-70} \doteq 0.8285$$

and then

$$A \doteq 34° \ ✓$$

EXERCISE SET L

Find the value of x in each triangle.

*1.

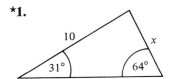

*2.

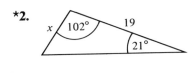

*Answers to starred problems are to be found at the back of the book.

3.

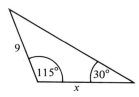

*4.

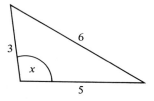

*5.

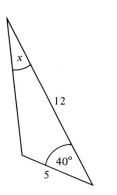

6.

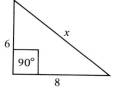

7.

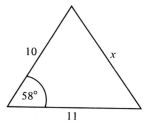

8.

chapter five

polynomial functions

Quadratic equations have appeared in several settings throughout the book. It is appropriate, then, to begin the study of polynomial functions with a look at one special kind of polynomial functions—quadratic functions.

5.1 QUADRATIC FUNCTIONS

Functions of the form $f(x) = ax^2 + bx + c$ $(a \neq 0)$ are called *quadratic functions*. The following are examples of quadratic functions:

$$f(x) = x^2 + 5x - 6$$
$$g(x) = 3x^2 - 5x + 2$$
$$h(x) = x^2 - 4$$
$$q(x) = x^2$$
$$y = x^2 + 6x$$

Associated with each quadratic function are two numbers called *zeros* of the function. They are the values of x for which the function is equal to zero. For example, if

$$f(x) = x^2 + 5x - 6$$

its zeros are the numbers for which

$$f(x) = 0$$

that is, for which $x^2 + 5x - 6 = 0$

The problem of finding the zeros of a quadratic function thus reduces to solving a corresponding quadratic equation. Such quadratic equations should be solved by factoring, if possible; otherwise, the quadratic formula can be used. The equation above is easily solved by factoring.

$$x^2 + 5x - 6 = 0$$
$$(x + 6)(x - 1) = 0$$
$$x + 6 = 0 \mid x - 1 = 0$$
$$x = -6 \quad \mid \quad x = 1$$

Thus, the *roots* of the *equation* are −6 and 1. The *zeros* of the *function* are −6 and 1. In general,

> The *zeros* of the *function* are the same as the *roots* of the *equation*.

If the function $f(x) = x^2 + 5x - 6$ is graphed, you will see that it crosses the x axis [that is, $f(x)$ is 0] at −6 and 1, the zeros of the function. The function is zero at −6 and 1.

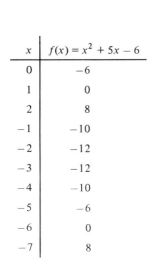

x	$f(x) = x^2 + 5x - 6$
0	−6
1	0
2	8
−1	−10
−2	−12
−3	−12
−4	−10
−5	−6
−6	0
−7	8

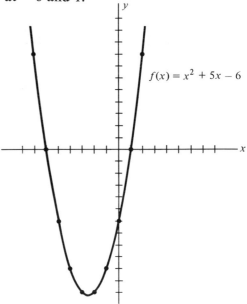

$f(x) = x^2 + 5x - 6$

If the zeros of a quadratic function are real and unequal, the curve will cross the x axis at two different places; the x values at both of those points represent zeros of the function. If the zeros of a quadratic function are real and equal (both the same), the tip of the curve will touch the x axis in one place and not cross it; the x value at that point is the "double" zero of the function. If the zeros are imaginary, the curve will not touch or cross the x axis. See the next figure for graphs of these three cases.

The Zeros of a Quadratic Function

(a) two different real zeros

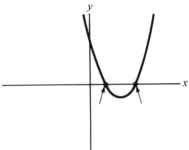

(b) one (double) real zero

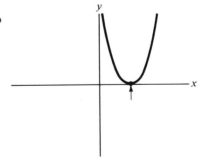

(c) two imaginary zeros

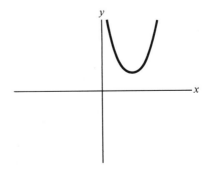

The nature of the zeros of a quadratic function can be determined by examining the *discriminant*, $b^2 - 4ac$. When the function is considered in the form $f(x) = ax^2 + bx + c$, the corresponding quadratic equation is $ax^2 + bx + c = 0$, and

$$x = \frac{-b \pm \sqrt{\boxed{b^2 - 4ac}}}{2a} \longleftarrow \text{discriminant}$$

If $b^2 - 4ac = 0$, then $x = (-b \pm 0)/(2a)$ or $x = -b/(2a)$. Both roots are the same. The roots here are real and equal. Thus, the zeros of the function are real and equal. If $b^2 - 4ac > 0$, then $\sqrt{b^2 - 4ac}$ is a positive number. The roots, then, are real and unequal. Thus, the zeros of the function are real and unequal. If $b^2 - 4ac < 0$, then $\sqrt{b^2 - 4ac}$ is an imaginary number. The roots, then, are imaginary and unequal. Thus, the zeros of the function are imaginary and unequal.

EXERCISE SET A

Find the zeros of each function in problems 1–13 below.

★1. $f(x) = x^2 + 7x + 12$ 2. $P(x) = x^2 - 9$

★3. $g(x) = x^2 + 16$ ★4. $p(x) = x^2 + x + 3$

5. $h(x) = 3x^2 + 7x - 1$ ★6. $d(x) = x^3 + 7x^2 + 10x$

7. $c(x) = x^2 - 5x - 14$ ★8. $f(x) = (x + 9)^2$

9. $c(x) = 3x^2 + 5x + 7$ 10. $d(x) = 5x^2 - 2x - 1$

★11. $t(x) = (x^2 + 6x + 8)(x^2 - 9)$

★12. $q(x) = x^4 - 1$ (the difference of two squares)

13. $r(x) = x^4 - 2x^2 - 8$ (quadratic in x^2)

★14. From the top of a 2400-foot cliff a stone is thrown upward with an initial velocity of 80 feet per second. Formulas for the velocity v (in feet per second) of the stone at any time t (in seconds) and the distance s (in feet) of the stone from the ground (that is, the bottom of the cliff) at any time t are readily derivable using calculus. Those formulas are:

$$\text{Velocity} = v(t) = -32t + 80$$
$$\text{Distance} = s(t) = -16t^2 + 80t + 2400$$

★Answers to starred problems are to be found at the back of the book.

Answer the following questions:

(a) How fast is the stone traveling 1 second after it is thrown?

(b) How far is the stone from the ground after 7 seconds?

(c) After how many seconds does the stone strike the ground?

(d) With what velocity does the stone strike the ground?

(e) How fast is the stone going when it passes the top of the cliff on the way down?

(f) How high (with respect to the ground) does the stone reach before it starts heading down toward the ground? (*Hint:* Think velocity as well as distance.)

15. Same as problem 14, except that $v(t) = -32t + 112$ and $s(t) = -16t^2 + 112t + 3168$.

16. Find the zeros of

(a) $f(x) = x^{1/3} - 2x^{4/3}$ (b) $g(x) = 2x^{1/3} - 2$

(c) $h(x) = \sin x$ (d) $k(x) = \ln x$

5.2 POLYNOMIALS

Quadratic functions are polynomial functions of degree 2. Here are several examples of polynomial functions. The *degree* of the polynomial, or highest power of the variable, is indicated.

$$f(x) = 5x^2 - 3x + 2 \qquad \text{(degree 2)}$$
$$g(x) = x^4 - 7x^3 - 4x^2 + 1 \qquad \text{(degree 4)}$$
$$h(x) = x^{20} - 5x^7 \qquad \text{(degree 20)}$$
$$j(x) = x^3 - \frac{7x^2}{9} + 4x - 3 \qquad \text{(degree 3)}$$
$$k(x) = x^2 - \sqrt{3}\,x + 9 \qquad \text{(degree 2)}$$

All of these fit the general form of a *polynomial function*:

$$P(x) = a_n x^n + a_{n-1} x^{n-1} + \cdots + a_2 x^2 + a_1 x^1 + a_0 x^0$$

where the numbers $a_0, a_1, a_2, \ldots, a_n$ are constants. Note that $x^0 = 1$, so $a_0 x^0 = a_0$ is the constant term. Also, $x^1 = x$.

Each term of a polynomial in x must be a constant times a whole-number power of x. Recall that the whole numbers are $0, 1, 2, 3, 4, \ldots$.

The following are not polynomials in x:

$$f(x) = x^5 - \frac{1}{x^2} + 4 \qquad \begin{cases} 1/x^2 = x^{-2}. \text{ The power of } x \\ \text{is } -2, \text{ not a whole number.} \end{cases}$$

$$g(x) = x^2 + 3\sqrt{x} + 4x - 9 \qquad \begin{cases} \sqrt{x} = x^{1/2}, \text{ and } \frac{1}{2} \text{ is not a} \\ \text{whole number.} \end{cases}$$

Polynomial functions of small degree are named. Those of first degree, $f(x) = ax + b$, are called linear polynomial functions or simply *linear functions*. Polynomials of degree two, $f(x) = ax^2 + bx + c$, are *quadratic functions*. Polynomials of degree three, $f(x) = ax^3 + bx^2 + cx + d$, are *cubic*. Polynomials of degree four are called *quartic*.

As suggested earlier, the zeros of a quadratic polynomial function are determined by factoring the quadratic expression and setting each factor equal to zero. However, if the expression cannot be factored, then the quadratic formula can be used to determine the zeros.

Factoring a cubic expression is more complicated than factoring a quadratic, since there are three factors instead of two. Expressions of still higher degree are even more complicated. The approach to factoring cubic expressions is to determine one factor first. When that factor is divided into the original cubic expression, the result is a quadratic expression, which can be factored as explained above. Here is an example of the long-division process:

$$x - 3\overline{)x^3 - 5x^2 + 7}$$

First, note that the x term is missing from the expression $x^3 - 5x^2 + 7$. Accordingly, insert $+0x$ in the expression to serve as a placeholder.

$$x - 3\overline{)x^3 - 5x^2 + 0x + 7}$$

Begin the division by dividing $x - 3$ into $x^3 - 5x^2$. This is done by dividing x into x^3. The result is x^2.

$$\begin{array}{r} x^2 \\ x - 3\overline{)x^3 - 5x^2 + 0x + 7} \end{array}$$

Now multiply the x^2 by $x - 3$ and subtract that product from $x^3 - 5x^2$. Then bring down the next term, $+0x$.

$$\begin{array}{r} x^2 \\ x - 3\overline{)x^3 - 5x^2 + 0x + 7} \\ \underline{x^3 - 3x^2} \downarrow \\ -2x^2 + 0x \end{array}$$

Now divide $x - 3$ into $-2x^2 + 0x$ by dividing x into $-2x^2$. The result is $-2x$.

$$\begin{array}{r} x^2 - 2x \\ x - 3\overline{)x^3 - 5x^2 + 0x + 7} \\ \underline{x^3 - 3x^2} \\ -2x^2 + 0x \end{array}$$

Multiply the $-2x$ by $x - 3$, subtract the product from $-2x^2 + 0x$, and then bring down the next term, $+7$.

$$\begin{array}{r} x^2 - 2x \\ x - 3\overline{)x^3 - 5x^2 + 0x + 7} \\ \underline{x^3 - 3x^2} \\ -2x^2 + 0x \\ \underline{-2x^2 + 6x} \\ -6x + 7 \end{array}$$

Finally, divide $x - 3$ into $-6x + 7$ by dividing x into $-6x$. The result is -6. Multiplying the -6 by $x - 3$ produces $-6x + 18$. A remainder of -11 is left after the subtraction.

$$\begin{array}{r} x^2 - 2x - 6 \\ x - 3\overline{)x^3 - 5x^2 + 0x + 7} \\ \underline{x^3 - 3x^2} \\ -2x^2 + 0x \\ \underline{-2x^2 + 6x} \\ -6x + 7 \\ \underline{-6x + 18} \\ -11 \quad \text{R} \quad \text{(remainder)} \end{array}$$

Since the division is complete and the remainder is not zero, we conclude that $x - 3$ is not a factor of $x^3 - 5x^2 + 7$.

Before trying to determine the zeros of a cubic polynomial function, note that a zero of a function is any number z such that $f(z) = 0$. Also, if z is a zero of polynomial function $f(x)$, then $(x - z)$ is a factor of the polynomial expression. For example, if $f(x) = x^2 + 5x - 6$, then -6 and 1 are zeros, since $f(-6) = 0$ and $f(1) = 0$. Also, $(x + 6)$ and $(x - 1)$ are factors.

$$x^2 + 5x - 6 = (x + 6)(x - 1)$$

$$-6 \quad \text{zero} \longrightarrow x + 6 \quad \text{factor}$$

$$+1 \quad \text{zero} \longrightarrow x - 1 \quad \text{factor}$$

In general, if z is a zero, then $(x - z)$ is a factor.

Example 1. *Find all the zeros of $f(x) = x^3 + 9x^2 + 17x + 6$.*

By trying different numbers for x, it is eventually determined that $f(-2) = 0$. Thus, -2 is a zero of $f(x)$, and $(x + 2)$ is a factor of the expression $x^3 + 9x^2 + 17x + 6$. When $x + 2$ is divided into the cubic expression, the result is a quadratic expression, $x^2 + 7x + 3$. Because $x + 2$ is a factor, there is no remainder in the division by $x + 2$.

$$
\begin{array}{r}
x^2 + 7x + 3 \\
x + 2 \overline{)x^3 + 9x^2 + 17x + 6} \\
\underline{x^3 + 2x^2} \\
7x^2 + 17x \\
\underline{7x^2 + 14x} \\
3x + 6 \\
\underline{3x + 6}
\end{array}
$$

The expression $x^2 + 7x + 3$ cannot be factored, but the roots of the equation $x^2 + 7x + 3 = 0$ can be obtained by using the quadratic formula.

$$x = \frac{-7 \pm \sqrt{49 - 12}}{2} = \frac{-7 \pm \sqrt{37}}{2}$$

So the zeros of the function are

$$-2 \quad \text{and} \quad \frac{-7 \pm \sqrt{37}}{2} \quad \checkmark$$

You are probably wondering how many trials it took to come up with -2 as a zero of the function, or how you should decide what numbers to test. Here is the key to the trial-and-error approach.

Theorem: *If a polynomial is of the form*

$$P(x) = x^n + a_{n-1}x^{n-1} + \cdots + a_1x + a_0$$

where $a_{n-1}, \ldots, a_1, a_0$ are integers, then the only possible rational zeros are the integer factors of the constant a_0.

In the polynomial above, $f(x) = x^3 + 9x^2 + 17x + 6$, the only possible rational zeros are $+6, -6, +3, -3, +2, -2, +1$, and -1. These numbers are the only integer factors of $+6$, so they are the only possible rational zeros of the polynomial. A quick look at the function shows that all connecting signs are positive. This means that $+6, +3,$ and $+1$ cannot be zeros. Why? There is no way that a sum of all positive numbers will be equal to zero. Thus, $-6, -3, -2,$ and -1 are the only numbers worth testing.

$$f(-6) = (-6)^3 + 9(-6)^2 + 17(-6) + 6 = \quad 12$$
$$f(-3) = (-3)^3 + 9(-3)^2 + 17(-3) + 6 = \quad 9$$
$$f(-2) = (-2)^3 + 9(-2)^2 + 17(-2) + 6 = \quad 0$$
$$f(-1) = (-1)^3 + 9(-1)^2 + 17(-1) + 6 = \quad -3$$

Only -2 is a zero; -2 is the only rational zero of the function. From this analysis we conclude that the other zeros are either irrational or imaginary. (In Example 1 they were found to be irrational.)

Example 2. *Find all the zeros of* $f(x) = x^3 + 4x^2 - 7x - 10.$

The only possible rational zeros are the factors of -10, namely, $\pm 10,$ $\pm 5, \pm 2, \pm 1.$ After a few tries we discover that $f(-5) = 0.$

$$f(-5) = (-5)^3 + 4(-5)^2 - 7(-5) - 10 = 0$$

Since -5 is a zero of the function, $x + 5$ is a factor of the expression $x^3 + 4x^2 - 7x - 10.$ After dividing the expression by $x + 5$, we obtain $x^2 - x - 2.$

$$
\begin{array}{r}
x^2 - x - 2 \\
x + 5{\overline{\smash{\big)}\,x^3 + 4x^2 - 7x - 10}} \\
\underline{x^3 + 5x^2} \\
-x^2 - 7x \\
\underline{-x^2 - 5x} \\
-2x - 10 \\
\underline{-2x - 10}
\end{array}
$$

The remaining quadratic expression can be factored as $x^2 - x - 2 = (x - 2)(x + 1)$, yielding 2 and -1 as the other zeros. Thus, the zeros of the function are $-5, 2,$ and $-1.$

If the original polynomial is of degree four, then division by the first factor obtained will reduce the quartic expression to cubic. Thus, another zero (and its associated factor) must be found and another division performed before a quadratic expression is produced.

If the coefficient of the highest-degree term is not 1, then the following theorem should be used to aid the search for zeros. It is an extension of the preceding theorem regarding rational zeros.

Theorem: *If a polynomial is of the form*

$$p(x) = a_n x^n + a_{n-1} x^{n-1} + \cdots + a_1 x + a_0$$

where a_n, a_{n-1}, ..., a_1, a_0 are integers, then the only possible rational zeros are the fractions formed using as numerators the factors of a_0 and as denominators the factors of a_n.

Example 3. *Find all the zeros of $f(x) = 2x^3 + x^2 + 5x - 3$.*

The only possible rational zeros have factors of -3 as their numerators and factors of 2 as their denominators. The possibilities are:

$$\pm \frac{3}{1}, \ \pm \frac{3}{2}, \ \pm \frac{1}{1}, \ \pm \frac{1}{2}$$

It is natural to try ± 1 first, and then ± 3 if neither $+1$ nor -1 is a zero. Unfortunately, all four of these numbers fail to be zeros. But $f(\frac{1}{2}) = 0$, so $\frac{1}{2}$ is a zero of the function. This means that $x - \frac{1}{2}$ is a factor of the expression. This also means that $2x - 1$ is a factor of the expression.

$$2x - 1 = 0$$

leads to

$$x = \frac{1}{2}$$

The factor can be written using integers before attempting to perform the division. (*If* $-\frac{5}{4}$ was a zero, then the factor $x + \frac{5}{4}$ can be written as $4x + 5$ for the division.) The division:

$$
\begin{array}{r}
x^2 + x + 3 \\
2x - 1 \overline{)2x^3 + x^2 + 5x - 3} \\
\underline{2x^3 - x^2} \\
2x^2 + 5x \\
\underline{2x^2 - x} \\
6x - 3 \\
\underline{6x - 3} \\
\end{array}
$$

The remaining quadratic expression cannot be factored. However, when the quadratic formula is applied to $x^2 + x + 3 = 0$, the result is

$$x = \frac{-1 \pm \sqrt{-11}}{2} = \frac{-1 \pm i\sqrt{11}}{2}$$

The zeros of the function are $\frac{1}{2}$, $(-1 + i\sqrt{11})/2$, and $(-1 - i\sqrt{11})/2$.

Perhaps you noticed that there were two complex zeros of the polynomial function above: $(-1 + i\sqrt{11})/2$ and $(-1 - i\sqrt{11})/2$. In general, *if $f(x)$ is a polynomial with real coefficients and if $a + bi$ is a zero of $f(x)$, then $a - bi$ is also a zero of $f(x)$.* Imaginary zeros always come in complex-conjugate pairs.

EXERCISE SET B

1. Perform each long division.
   *(a) Divide $x^3 + 2x^2 - 13x - 6$ by $x - 3$
   *(b) Divide $x^3 - x^2 - 25x + 25$ by $x - 5$
   (c) Divide $x^3 + 3x^2 - 2x - 3$ by $x - 2$
   (d) Divide $x^3 + 7x^2 + 13x + 6$ by $x + 2$
   (e) Divide $x^3 + 3x^2 - 5x - 39$ by $x + 3$
   *(f) Divide $x^3 - 2x^2 + 4x + 1$ by $x + 2$
   *(g) Divide $2x^4 - 3x^2 - 7x + 3$ by $x - 2$
   (h) Divide $2x^4 + x^3 - 2x - 1$ by $x + \frac{1}{2}$
   (i) Divide $2x^4 + x^3 - 2x - 1$ by $2x + 1$
   *(j) Divide $x^4 - 2x + 1$ by $x + 2$
   *(k) Divide $x^3 - 1$ by $x - 1$
   *(l) Divide $x^3 - y^3$ by $x - y$
   (m) Divide $x^3 + y^3$ by $x + y$

2. Factor each cubic polynomial expression completely.
   *(a) $x^3 - 4x^2 + x + 6$ *(b) $x^3 - 9x^2 + 20x$
   *(c) $x^3 + 4x^2 + x - 6$ (d) $x^3 + x^2 - 16x - 16$
   (e) $x^3 + 11x^2 + 39x + 45$ (f) $x^3 - x^2 - x + 1$
   *(g) $2x^3 + 19x^2 + 49x + 20$ (h) $2x^3 - 5x^2 - x + 6$
   *(i) $3x^3 + 11x^2 - 14x - 40$ (j) $6x^3 + 29x^2 - 7x - 10$

3. Determine all the zeros of each polynomial function.
   *(a) $f(x) = x^3 - 5x^2 + x - 5$

*Answers to starred problems are to be found at the back of the book.

(b) $p(x) = x^3 - x^2 - 9x + 9$

*(c) $t(x) = x^3 + 8$

(d) $g(x) = x^3 - 7x - 6$

*(e) $f(x) = x^3 + 4x^2 - 2x - 20$

(f) $q(x) = x^3 + 8x^2 + 14x - 5$

*(g) $f(x) = x^3 + 7x^2 + 15x + 25$

*(h) $g(x) = 2x^3 + 5x^2 - 4x - 3$

*(i) $h(x) = x^4 - 5x^3 - 13x^2 - 5x - 14$

(j) $r(x) = x^4 + 3x^3 + 11x^2 + 27x + 18$

(k) $s(x) = 4x^3 - 11x^2 + x + 1$

*(l) $f(x) = 2x^4 - x^3 + x^2 - 41x - 21$

(m) $m(x) = x^4 - 16$

4. Find all roots of the following polynomial equations.

*(a) $x^3 - 4x^2 + x + 6 = 0$ *(b) $x^3 - 3x^2 + 4x - 12 = 0$

(c) $x^3 + 7x^2 + 13x + 6 = 0$ *(d) $2x^4 + x^3 - 2x - 1 = 0$

(e) $2x^3 + x^2 + 14x + 7 = 0$

5. Show that each polynomial function has no rational zeros.

(a) $f(x) = x^3 + 2$ (b) $g(x) = x^3 + 2x^2 + 5x + 1$

(c) $h(x) = x^4 + x^2 + 2x + 6$ (d) $n(x) = x^4 - x + 17$

(e) $t(x) = x^5 + x^3 + 9$

6. Determine the three cube roots of -27 by solving the polynomial equation $x^3 = -27$.

7. Solve each inequality. Factoring should help.

*(a) $x^3 - 4x^2 - x + 4 > 0$ (b) $x^3 - 2x^2 - 5x + 6 < 0$

5.3 GRAPHS OF POLYNOMIALS

The graph of a polynomial function is a smooth continuous curve. To sketch the graph of a polynomial, obtain and plot enough points to give the general shape of the curve. Then draw a smooth continuous curve through the points. Real zeros of the function occur where the graph crosses the x axis. Exact values of rational zeros can be obtained by earlier techniques. Irrational zeros of polynomials can be located by applying the following theorem for polynomials.

Theorem: *If a and b are real numbers for which f (a) and f (b) have opposite signs, then there is at least one real number c between a and b such that f (c) = 0.*

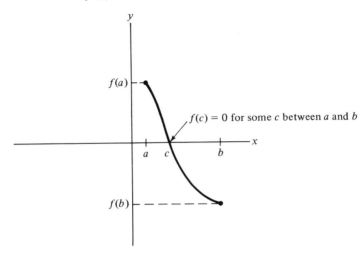

Example 4. *Sketch the graph of* $f(x) = x^3 - 5x^2 + 2x + 1$.

First, obtain some points.

$$f(0) = +1 \rightarrow (0, 1)$$
$$f(1) = -1 \rightarrow (1, -1)$$

There is a real zero between 0 and 1, because $f(0)$ is positive and $f(1)$ is negative. Continuing,

$$f(2) = -7 \ \rightarrow (2, -7)$$
$$f(3) = -11 \rightarrow (3, -11)$$
$$f(4) = -7 \ \rightarrow (4, -7)$$
$$f(5) = +11 \rightarrow (5, 11)$$

There is a real zero between 4 and 5, since $f(4)$ is negative and $f(5)$ is positive. Continuing,

$$f(-1) = -7 \rightarrow (-1, -7)$$

There is a real zero between -1 and 0, since $f(-1)$ is negative and $f(0)$ is positive.

From earlier theory, the only possible rational zeros are factors of 1, namely, $+1$ and -1. But $f(1) = -1$ and $f(-1) = -7$. So there are no rational zeros. This means that the three zeros located above are irrational. A sketch of the curve is as follows:

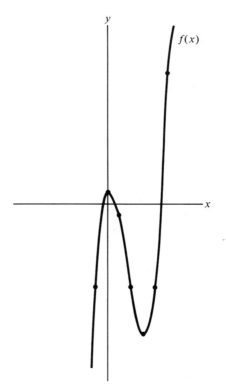

5.4 THEOREMS

There are several relevant theorems listed below. In each case it is assumed that the degree of the polynomial is at least one.

Theorem 1 (The Remainder Theorem): *The remainder in the division of a polynomial $p(x)$ by $x - r$ is $p(r)$.*

Theorem 2 (Factor Theorem): *If $x - r$ is a factor of polynomial $p(x)$, then $p(r) = 0$. Also, if $p(r) = 0$, then $x - r$ is a factor of $p(x)$.*

Theorem 3 (Fundamental Theorem of Algebra): *If $p(x)$ is a polynomial, then there is at least one number a such that $p(a) = 0$.*

Theorem 4: *A polynomial of degree n has exactly n zeros (not necessarily distinct).*

EXERCISE SET C

1. Sketch the graph of each polynomial function. All zeros are rational.
   (a) $f(x) = x^3 - x^2 - 2x$ (b) $g(x) = x^3 - 7x - 6$
   (c) $f(x) = x^3 - x^2 - x + 1$ (d) $f(x) = x^3 + x^2 - 16x - 16$
   (e) $h(x) = (x - 2)(x - 4)(x + 1)$ (f) $g(x) = x(x - 1)(x + 2)(x + 1)$

2. Sketch the graph of each polynomial function. Show the location of all real zeros (between consecutive integers).
   (a) $f(x) = x^3 - 2x^2 - 3x + 5$ (b) $p(x) = x^3 + x^2 - 2x - 2$
   (c) $g(x) = -x^3 + 3x^2 - x - 4$ (d) $h(x) = x^3 - 3x^2 + 2x + 1$
   (e) $t(x) = x^3 + 3x^2 - 2x - 6$ (f) $m(x) = -x^3 + 4x^2 - 4$

NOTE. The *Study Guide* contains a supplementary chapter on synthetic division and theory of polynomials.

chapter six

---

coordinate geometry

In this chapter we will study classes of elementary curves and the algebraic equations which describe them. The study begins with straight lines.

The equation of any straight line can be written in the form

$$ax + by + c = 0$$

in which a and b cannot both be zero simultaneously. Equations of lines are not always written this way, nor is this always a useful form in which to write such equations. But any line *can* be written this way. All lines fit this form. Thus, changing a relation to fit this form will determine whether or not it is a line. Here are some equations of lines.

$$15x + 3y + 7 = 0$$
$$2x + y = 6$$

$$y = 9x + 14$$
$$x = -8$$
$$y - 12 = 0$$

The first line, $15x + 3y + 7 = 0$, is already in the standard form. The second line, $2x + y = 6$, can be changed to standard form by adding -6 to both sides of the equation. The result is $2x + y - 6 = 0$. The third line, $y = 9x + 14$, can be changed to standard form by adding $-9x$ and -14 to both sides of the equation. The result is $-9x + y - 14 = 0$ or $9x - y + 14 = 0$. The line $x = -8$ is changed by adding 8 to both sides in order to obtain $x + 8 = 0$. The equation $x + 8 = 0$ is in the standard form $ax + by + c = 0$; b is zero. The line $y - 12 = 0$ is already in standard form; a is zero.

The inclination or slant of a line with respect to the x axis is called the *slope* of the line. It is measured by comparing the change in the y coordinates with the change in the x coordinates for any two points on the line. Specifically,

$$\text{Slope} = \frac{\text{change in } y}{\text{change in } x}$$

Using the letter m for slope and the symbol Δ (delta) for change in, this becomes

$$m = \frac{\Delta y}{\Delta x}$$

Consider the line $y = 2x + 3$. Here are two points of the line:

x	y	
0	3	$\longrightarrow (0, 3)$
1	5	$\longrightarrow (1, 5)$

So

$$m = \frac{\Delta y}{\Delta x} = \frac{5 - 3}{1 - 0} = \frac{2}{1} = 2 \quad \checkmark$$

$m = 2$

The slope of the line is 2. To avoid getting the wrong sign when computing slope, it is necessary to be consistent when determining the difference in the y's and the difference in the x's. Subtract both coordinates of one point from the corresponding coordinates of the other point. If (x_1, y_1) represents one point and (x_2, y_2) represents the other, then

TWO POINTS

$$m = \frac{\Delta y}{\Delta x} = \frac{y_2 - y_1}{x_2 - x_1}$$

Example 1. *Find the slope of the line which passes through the points* (3, 7) *and* (5, 4).

It does not matter which point is called (x_1, y_1), so let (3, 7) be (x_1, y_1). Then (5, 4) is (x_2, y_2). Now the slope is computed as

$$m = \frac{\Delta y}{\Delta x} = \frac{y_2 - y_1}{x_2 - x_1} = \frac{4 - 7}{5 - 3} = \frac{-3}{2} = -\frac{3}{2} \quad \checkmark$$

The slope is $-\frac{3}{2}$, which demonstrates that slopes can be negative.

Let's draw the two lines discussed above to see what different slopes look like.

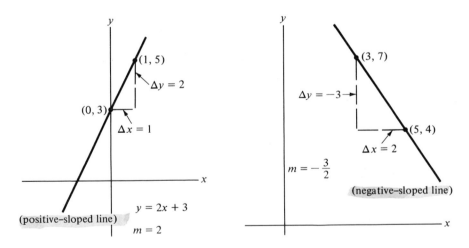

Since the slope of the first line is 2, or $\frac{2}{1}$, another point on that line can be obtained by starting at (1, 5) and changing x by 1 and y by 2. That point is (2, 7).   Next page

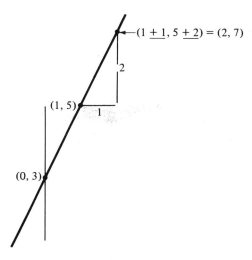

Similarly, since the slope of the second line is $-\frac{3}{2}$, another point on that line can be determined by starting at (5, 4) and changing x by 2 and y by -3. That point is (7, 1). $Below$

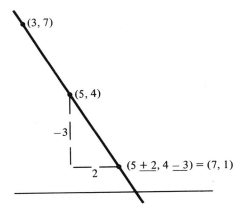

Recall that the slope, m, of the line through points (x_1, y_1) and (x_2, y_2) can be computed as

$$m = \frac{y_2 - y_1}{x_2 - x_1}$$

ONE

If instead of the specific point (x_2, y_2) we use any point (x, y), the result is

$$m = \frac{y - y_1}{x - x_1}$$

If a specific value is supplied for slope m, then the preceding equation becomes a relation between x and y. In fact, it is a linear relation. Specifically, it is the equation of the line with slope m which passes through the point (x_1, y_1).

Example 2. *Find the equation of the line with slope 3 which passes through the point* (2, 4).

Substitute 3 for m and (2, 4) for (x_1, y_1) into

$$m = \frac{y - y_1}{x - x_1}$$

to obtain

$$3 = \frac{y - 4}{x - 2}$$

which can be simplified. First multiply both sides by $x - 2$.

$$3(x - 2) = y - 4$$

Then

$$3x - 6 = y - 4$$

or

$$y = 3x - 2 \quad \checkmark$$

or

$$3x - y - 2 = 0 \quad \checkmark$$

Perhaps you realize that each time such a problem is solved, it is necessary to multiply by $x - x_1$ in order to simplify the relation. Because of this the formula

$$m = \frac{y - y_1}{x - x_1}$$

is usually written

$$\boxed{y - y_1 = m(x - x_1)}$$

This is the traditional form of the *point-slope formula*.

Example 3. *Find the equation of the line with slope $\frac{7}{2}$ which passes through the point $(-3, 1)$.*

Use

$$y - y_1 = m(x - x_1)$$

with

$$m = \frac{7}{2}$$

$$x_1 = -3$$

$$y_1 = 1$$

The result is

$$y - 1 = \tfrac{7}{2}(x - -3)$$

or

$$y - 1 = \tfrac{7}{2}(x + 3)$$

Now multiply both sides by 2 to eliminate the fraction.

$$2y - 2 = 7(x + 3)$$
$$2y - 2 = 7x + 21$$
$$7x - 2y + 23 = 0 \quad \checkmark$$

Example 4. *Determine the equation of the line which passes through the points $(1, 6)$ and $(3, 0)$.*

The slope of the line is computed as

$$m = \frac{y_2 - y_1}{x_2 - x_1} = \frac{0 - 6}{3 - 1} = \frac{-6}{2} = -3$$

Now use -3 for m and either point that was given, say $(3, 0)$, in the same way they were used in the preceding example.

$$y - y_1 = m(x - x_1)$$

becomes

$$y - 0 = -3(x - 3)$$

or

$$y = -3x + 9 \quad \checkmark$$

This can also be written as

$$3x + y - 9 = 0 \quad \checkmark$$

The slope of a line has been defined as $\Delta y/\Delta x$. This is the same as the tangent of the angle of inclination α of the line with respect to the x axis.

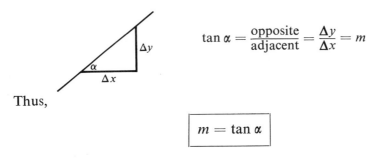

$$\tan \alpha = \frac{\text{opposite}}{\text{adjacent}} = \frac{\Delta y}{\Delta x} = m$$

Thus,

$$\boxed{m = \tan \alpha}$$

Parallel lines have the same slope. They do not meet because they are slanted or sloped the same. The angle of inclination α is the same for two lines that are parallel. The lines $y = 2x + 5$ and $y = 2x - 3$ are parallel, while $y = 2x + 5$ and $y = 5x + 2$ are not parallel.

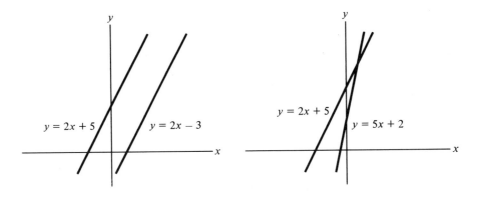

There is a relationship between the slopes of two *perpendicular* lines. If lines l_1 and l_2 are perpendicular and have slopes m_1 and m_2, respectively, then

$$\boxed{m_2 = -\frac{1}{m_1}}$$

To prove this, consider the next illustration.

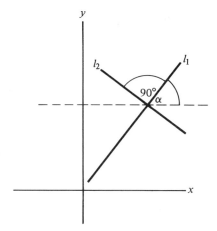

The slope m_1 of line 1 is $\tan \alpha$. The slope m_2 of line 2 is $\tan(\alpha + 90°)$. Thus,

$$m_2 = \tan(\alpha + 90°)$$

$$= -\cot \alpha \qquad \left\{ \begin{array}{l} \text{by problem 3, Exercise} \\ \text{Set F, Chapter 4} \end{array} \right.$$

$$= -\frac{1}{\tan \alpha} \qquad \left(\text{since } \cot \alpha = \frac{1}{\tan \alpha} \right)$$

$$= -\frac{1}{m_1} \qquad (\text{since } m_1 = \tan \alpha)$$

Example 5. *Determine the equation of the line which passes through the point* $(1, -3)$ *and is perpendicular to the line* $2x + y - 5 = 0$.

Obtain two points in order to compute the slope of the line.† If $x = 0$, $y = 5$. If $x = 1$, $y = 3$. Thus, two points on the line are $(0, 5)$ and $(1, 3)$. The slope of the line is

$$m = \frac{\Delta y}{\Delta x} = \frac{5 - 3}{0 - 1} = \frac{2}{-1} = -2$$

The slope of the line perpendicular to $2x + y - 5 = 0$ is $\frac{1}{2}$, since

$$m_\perp = -\frac{1}{m} = -\frac{1}{-2} = \frac{1}{2} \qquad \left\{ \begin{array}{l} \text{The symbol } \perp \text{ means} \\ \text{perpendicular.} \end{array} \right.$$

†In the next section you will learn how to determine the slope of a line from its equation—without obtaining any points.

So we now seek the equation of the line through $(1, -3)$ with slope $\frac{1}{2}$. Using

$$y - y_1 = m(x - x_1)$$

we get

$$y - -3 = \tfrac{1}{2}(x - 1)$$
$$y + 3 = \tfrac{1}{2}(x - 1)$$
$$2y + 6 = x - 1 \qquad \left\{ \begin{array}{l} \text{if both sides are multiplied} \\ \text{by 2} \end{array} \right.$$

or

$$x - 2y - 7 = 0 \quad \checkmark$$

The *midpoint* of a line segment is the point on the segment halfway between the endpoints of the segment. This means that the x coordinate of the midpoint is the average of the x coordinates of the endpoints. Similarly, the y coordinate of the midpoint is the average of the y coordinates of the endpoints. Thus, if (x_1, y_1) and (x_2, y_2) are the endpoints of a line segment, then the midpoint of the segment, $(\bar{x}, \bar{y})$, is given by

$$(\bar{x}, \bar{y}) = \left(\frac{x_1 + x_2}{2}, \frac{y_1 + y_2}{2} \right)$$

If this point $(\bar{x}, \bar{y})$ is indeed the midpoint of the segment joining (x_1, y_1) and (x_2, y_2), then the distance from (x_1, y_1) to $(\bar{x}, \bar{y})$ must equal the distance from $(\bar{x}, \bar{y})$ to (x_2, y_2). It is left as an exercise to prove that those two distances are equal, once a distance formula is derived (Section 6.3).

Example 6. *Find the midpoint of the line segment joining the points* $(4, 7)$ *and* $(8, 10)$.

$$\text{midpoint} = (\bar{x}, \bar{y}) = \left(\frac{4 + 8}{2}, \frac{7 + 10}{2} \right) = (6, 8\tfrac{1}{2}) \quad \checkmark$$

Example 7. *Determine the equation of the line which is the perpendicular bisector of the segment connecting* $(3, -4)$ *and* $(9, 10)$.

The perpendicular bisector must bisect the segment connecting $(3, -4)$ and $(9, 10)$. That is, it must pass through the midpoint of that segment. And the midpoint of the segment is $(6, 3)$.

The perpendicular bisector must also be perpendicular to the line passing the points $(3, -4)$ and $(9, 10)$. Its slope, then, is the negative of the reciprocal of the slope of the line through $(3, -4)$ and $(9, 10)$.

$$m_{\text{segment}} = \frac{10 - (-4)}{9 - (3)} = \frac{14}{6} = \frac{7}{3}$$

$$m_{\perp} = -\frac{1}{\frac{7}{3}} = -\frac{3}{7}$$

So the perpendicular bisector of the original segment has a slope of $-\frac{3}{7}$ and passes through the point $(6, 3)$. The equation of the perpendicular bisector can now be readily obtained from the formula $y - y_1 = m(x - x_1)$. The equation becomes

$$y - 3 = -\tfrac{3}{7}(x - 6)$$

which can be simplified by multiplying both sides by 7.

$$7y - 21 = -3x + 18$$

or

$$3x + 7y - 39 = 0 \quad \checkmark$$

6.2 SLOPES, INTERCEPTS, AND LINEAR FUNCTIONS

Consider the line $y = 3x - 2$.

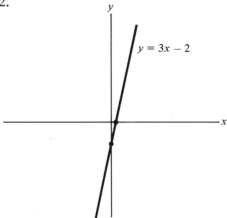

The only points marked are those where the line crosses an axis.

A line crosses the y axis where $x = 0$; that is, $x = 0$ where you are zero units from the y axis. When you are on the y axis, $x = 0$. The value of y when $x = 0$ (that is, where the line crosses the y axis) is called the y *intercept*. The y intercept for $y = 3x - 2$ is -2, since when $x = 0$ $y = 3 \cdot 0 - 2 = 0 - 2 = -2$.

The y intercept for the line $y = 7x + 9$ is 9, since $y = 9$ when $x = 0$.

Note that if lines are written in the form $y = ax + b$, then ax disappears (is equal to zero) when $x = 0$, so b is always the y intercept. Previously, lines were described by $y - y_1 = m(x - x_1)$. All lines except those parallel to the y axis must cross the y axis. Thus, all such lines have a y intercept, and it occurs at the point $(0, b)$. If $(0, b)$ is substituted for (x_1, y_1) in $y - y_1 = m(x - x_1)$, the result is

$$y - b = m(x - 0)$$

or

$$y - b = mx$$

or

$$y = mx + b$$

This shows that if a line is written in the form $y = mx + b$, then m is the slope and b is the y intercept. The form $y = mx + b$ is called the *slope-intercept* form of a line.

$$y = mx + b$$
$$m = \text{slope}$$
$$b = y \text{ intercept}$$

Example 8. *Find the slope and y intercept of* $y = 12x + 5$.

$y = 12x + 5$ is already in $y = mx + b$ form
$$m = 12 \quad \text{(slope)} \quad \checkmark$$
$$b = 5 \quad \text{(}y \text{ intercept)} \quad \checkmark$$

Example 9. *Find the slope and y intercept of* $2y - 3x = -5$.

The line $2y - 3x = -5$ is not in $y = mx + b$ form, so you cannot read off the slope and y intercept. Change it to $y = mx + b$ (slope-intercept) form.

Add $3x$ to both sides to isolate the y term.

$$2y - 3x = -5$$
$$\underline{ 3x \qquad 3x}$$
$$2y \qquad = \quad 3x - 5$$

Next, multiply by $\frac{1}{2}$ to get $y = mx + b$ form. The result is

$$y = \frac{1}{2} \cdot (3x - 5)$$

$$y = \frac{3}{2}x - \frac{5}{2}$$

Clearly,

$$m = \frac{3}{2} \qquad \text{(slope)} \quad \checkmark$$

$$b = -\frac{5}{2} \qquad (y \text{ intercept}) \quad \checkmark$$

Example 10. *Graph the line with slope* 2 *and y intercept* 1.

We are given $m = 2$ and $b = 1$. The fact that $b = 1$ means that $(0, 1)$ is on the line, since the y intercept, b, is the value of y when $x = 0$. Also, $m = 2 = 2/1 = \Delta y/\Delta x$.

If we use $(0, 1)$ as a first point and move from it by changing x by 1 and y by 2 (because $\Delta y/\Delta x = 2/1$), then we get $(1, 3)$ as another point on the line. The line can now be drawn, since we have two points.

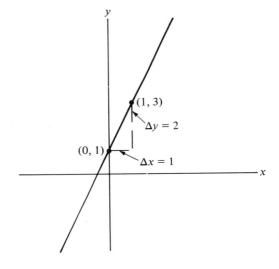

Also of some interest is the x intercept. The x *intercept* is the value of x when y is 0. Thus, the x intercept is the x value where the line crosses the x axis.

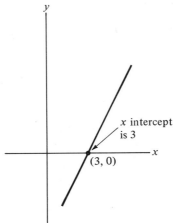

Example 11. *Find the x intercept of* $y = 5x - 1$.

To get the x intercept, let $y = 0$ in $y = 5x - 1$.

$$0 = 5x - 1$$
$$5x = 1$$
$$x = \frac{1}{5} \quad \checkmark$$

The x intercept is $\frac{1}{5}$.

As noted earlier, all lines except those parallel to the y axis can be written in the form $y = mx + b$. This means that all lines that are functions can be written as $y = mx + b$. Using function notation, the form becomes

$$f(x) = mx + b$$

Functions of the form $f(x) = mx + b$ are called *linear functions*.

EXERCISE SET A

1. Change each line to the form $ax + by + c = 0$.
   *(a) $4x + 3y = -5$ (b) $4x - 8 = 7y$

   *Answers to starred problems are to be found at the back of the book.

⋆(c) $7y + 2 = x$ (d) $5x = 2y + 15$
⋆(e) $x = 4 - 9y$ (f) $y = 17$
⋆(g) $3x = 1$ (h) $3y - 2x = 9$
⋆(i) $y = 6x - 4$ (j) $y = 8 - 3x$

2. Find the slope of the line through each pair of points.
   ⋆(a) (5, 7) and (3, 6)   (b) (10, 8) and (6, 4)
   ⋆(c) (4,2) and (6, 8)   (d) (5, 1) and (6, 0)
   ⋆(e) (0, 0) and (1, 5)   (f) (7, 0) and (4, 3)
   ⋆(g) (2, −3) and (−7, 4)   (h) (−5, 3) and (−3, −1)

⋆3. Find the equation of the line through each pair of points in problem 2.

⋆4. What is the slope of a line inclined at an angle of 60° with respect to the x axis?

5. Which of the following points are on the line $y = 4x - 3$?
   (a) (0, −3)   (b) (1, −1)   (c) (−1, 1)
   (d) (0, 4/3)   (e) (−11, −2)   (f) (2, 5)

⋆6. What is the equation of the x axis, and what is its slope?

7. What is the equation of the y axis, and what is its slope?

⋆8. Write the equation of the line that passes through the point (5, 3) and is parallel to the line $y = 2x + 3$.

9. Find the equation of the line passing through the point (3, −1) and perpendicular to the line $y = 6 - 2x$.

⋆10. Determine the midpoint of the segment connecting each pair of points.
   (a) (1, 4), (3, 10)   (b) (−2, 6), (4, 0)
   (c) (−3, 1), (2, −7)   (d) (0, 0), (−8, 7)
   (e) (−5, 0), (0, 3)

⋆11. Write the equation of the line that is the perpendicular bisector of the segment connecting the points.
   (a) (1, 2) and (5, 10)   (b) (2, 3) and (4, 4)
   (c) (1, 2) and (4, 5)   (d) (6, 3) and (4, 5)
   (e) (0, 8) and (−3, 9)   (f) (0, 0) and (7, 0)

12. Write each relation in $y = mx + b$ form, and determine the slope (m) and y intercept (b).
   ⋆(a) $y = 2x - 1$ ⋆(b) $y = -x + 4$

*(c) $y - 2 = x$ (d) $y = -3x + 2$
*(e) $y = 3(x + 1)$ *(f) $x + y = 0$
*(g) $2x - y = 6$ (h) $2x + 5y = 6$

*(i) $x = 3y - 4$ (j) $x = \dfrac{y}{2}$

*(k) $y - 2x + 3 = 0$ (l) $4x = 2y + 8$
(m) $2x = 3y + 18$ (n) $3x + y = 5$
*(o) $3(x + y) - 1 = x$ (p) $y = 4x + 5y - 8$

13. Show that if a line is of the form $y = mx + b$, then the x intercept is $-b/m$.

14. Graph the lines described by points, slope, and/or intercept.
    (a) $(2, 5)$, $(3, 8)$ (b) $(1, 0)$, $(0, -1)$
    (c) $m = 2$, $(2, 5)$ (d) $m = 1$, $(1, 4)$
    (e) $m = -3$, $(1, 0)$ (f) $m = 3$, $b = 2$
    (g) $m = 5$, $b = -1$ (h) $b = 5$, $(2, 3)$
    (i) $b = 2$, x intercept $= 1$ (j) $m = 3$, x intercept $= -2$

*15. Determine the equation of the line with x intercept -3 and parallel to the y axis.

16. Write the equation of the line with y intercept 4 and parallel to the x axis.

17. Determine the equation of the line with x intercept 5 and parallel to the line $x + 3y - 7 = 0$.

18. Write the equation of a line perpendicular to the line $3x + 5y - 6 = 0$.

*19. Write the equation of the line which has a y intercept of 6 and an x intercept of 2.

20. Prove that the line

$$\frac{x}{a} + \frac{y}{b} = 1$$

has x intercept equal to a and y intercept equal to b.

21. Write the equation of the line with slope p/q and y intercept r. Be sure to show the line in $y = mx + b$ form and in $ax + by + c = 0$ form.

6.3 DISTANCES

Now we shall derive a formula for the distance between any two points (x_1, y_1) and (x_2, y_2).

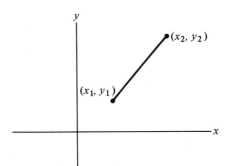

Consider the right triangle which includes these two points as vertices.

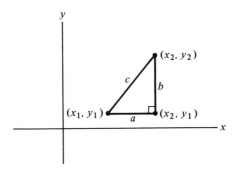

The sides are labeled a, b, and c. We seek a formula for c, the distance between (x_1, y_1) and (x_2, y_2). By the Pythagorean theorem,

$$c^2 = a^2 + b^2$$

In this setting,

$$a = |x_2 - x_1|$$

and

$$b = |y_2 - y_1|$$

so

$$c^2 = |x_2 - x_1|^2 + |y_2 - y_1|^2 = (x_2 - x_1)^2 + (y_2 - y_1)^2$$

or

$$c = \sqrt{(x_2 - x_1)^2 + (y_2 - y_1)^2}$$

is the *distance between the points* (x_1, y_1) *and* (x_2, y_2). As an example, the distance between the points $(2, 3)$ and $(6, 10)$ is equal to

$$\sqrt{(6 - 2)^2 + (10 - 3)^2}$$

which simplifies to $\sqrt{16 + 49}$ or $\sqrt{65}$.

The *distance from a point* (x_1, y_1) *to a line* $ax + by + c = 0$ is

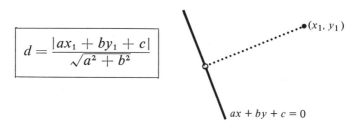

$$d = \frac{|ax_1 + by_1 + c|}{\sqrt{a^2 + b^2}}$$

$ax + by + c = 0$

The distance referred to here is the length of the segment beginning at (x_1, y_1), ending on the line $ax + by + c = 0$, and perpendicular to that line. Note that the line must be written in the form $ax + by + c = 0$ in order to obtain the correct numbers to use for a, b, and c.

The formula was obtained by the process described next. First, the slope of the line $ax + by + c = 0$ was obtained. Then the slope m_1 of a line perpendicular to it was determined. Next, the equation of the line through (x_1, y_1) with slope m_1 was determined. Then, the point of intersection (x_2, y_2) of this new line with $ax + by + c = 0$ was determined. Finally, the length of the segment between (x_1, y_1) and (x_2, y_2) was computed.

Example 12. *Find the distance between the point* $(5, -7)$ *and the line* $y = 2x - 3$.

The line $y = 2x - 3$ must be changed to $ax + by + c = 0$ form. This can be done by adding $-2x$ and $+3$ to both sides of the equation. The equation becomes $-2x + 1y + 3 = 0$, and the point is $(5, -7)$. So

$$d = \frac{|ax_1 + by_1 + c|}{\sqrt{a^2 + b^2}}$$

$$= \frac{|(-2)(5) + (1)(-7) + 3|}{\sqrt{(-2)^2 + (1)^2}}$$

$$= \frac{|-10 - 7 + 3|}{\sqrt{4 + 1}}$$

$$= \frac{14}{\sqrt{5}}$$

$\begin{cases} a = -2 \\ b = 1 \\ c = 3 \\ x_1 = 5 \\ y_1 = -7 \end{cases}$

The distance is $14/\sqrt{5}$, or approximately 6.26.

EXERCISE SET B

1. Find the distance between each pair of points.
  ★(a) $(1, 1)$ and $(5, 4)$ ★(b) $(4, 0)$ and $(4, 12)$
  ★(c) $(8, 2)$ and $(10, 7)$ (d) $(5, -2)$ and $(-3, 2)$

2. Compute the distance between the given point and the line.
  ★(a) $(2, 1)$, $3x + 4y + 7 = 0$ (b) $(4, 6)$, $2x + 3y - 5 = 0$
  ★(c) $(3, 4)$, $5x - 2y = 1$ (d) $(2, -3)$, $x - 6y = 7$
  ★(e) $(7, 0)$, $y = 2x + 3$ (f) $(1, -3)$, $y = -5x + 2$
  ★(g) $(-1, -5)$, $x - y = 9$ (h) $(9, -8)$, $x = 3y - 2$

3. Let $(\bar{x}, \bar{y}) = \left(\dfrac{x_1 + x_2}{2}, \dfrac{y_1 + y_2}{2} \right)$

Show that the distance from (x_1, y_1) to $(\bar{x}, \bar{y})$ equals the distance from $(\bar{x}, \bar{y})$ to (x_2, y_2).

6.4 CONIC SECTIONS

We now begin a study of curves described by second degree equations. Four kinds of curves are studied in this section: circles, parabolas, ellipses, and hyperbolas. Each of these curves is an example of a *conic section* (or less formally, a *conic*). Each type of curve can be obtained by the intersection of a plane with a right circular cone (hence the word "conic"). The nature of the intersecting plane determines which curve results.

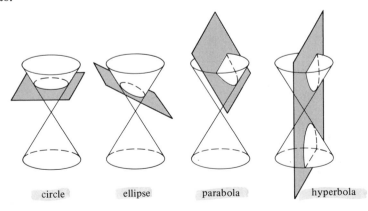

circle   ellipse   parabola   hyperbola

★Answers to starred problems are to be found at the back of the book.

A *circle* is a set of points whose distance from a fixed point is constant. The fixed point is called the *center.* The constant distance is called the *radius.* Thus, for a circle with radius r, center (h, k), and points of the form (x, y), application of the distance formula yields

$$\sqrt{(x - h)^2 + (y - k)^2} = r$$

If both sides are squared, the relation becomes

$$(x - h)^2 + (y - k)^2 = r^2$$

which is one standard form of a circle with center at (h, k) and radius of length r.

Example 13. *Write the equation of the circle with center at $(5, -3)$ and radius 9.*

The circle:
$$(x - 5)^2 + (y + 3)^2 = 81 \quad \checkmark$$

Example 14. *Write the equation of the circle with center at the origin and radius 5.*

With center at $(0, 0)$ and radius 5, we have
$$(x - 0)^2 + (y - 0)^2 = 5^2$$
or
$$x^2 + y^2 = 25 \quad \checkmark$$

In general, a circle with center at the origin and radius r has the form

$$x^2 + y^2 = r^2$$

Example 15. *Find the center and radius of the circle*
$$x^2 + y^2 + 8x - 10y - 8 = 0.$$

To find the center of the circle and radius, change it to the standard form,
$$(x - h)^2 + (y - k)^2 = r^2$$

This can be done by grouping the x terms together and the y terms together and then completing the two squares. Begin with

$$x^2 + y^2 + 8x - 10y - 8 = 0$$

First group the terms as suggested.

$$(x^2 + 8x) + (y^2 - 10y) = 8$$

Completing the square for $x^2 + 8x$ yields $(x + 4)^2$ or $x^2 + 8x + 16$. So 16 must be added to the right side of the equation to compensate for the additional 16 on the left side.

$$(x + 4)^2 + (y^2 - 10y) = 8 + 16$$

Similarly, completing the square for $y^2 - 10y$ yields $(y - 5)^2$, or $y^2 - 10y + 25$. So 25 must be added to the right side of the equation. The result is

$$(x + 4)^2 + (y - 5)^2 = 8 + 16 + 25$$

or

$$(x + 4)^2 + (y - 5)^2 = 49$$

So the center is at $(-4, 5)$ and the radius is 7.

A *parabola* is the set of points whose distances from a fixed point (called the *focus*) and a fixed line (called the *directrix*) are equal. The point of the parabola that lies on the perpendicular from the focus to the directrix is called the *vertex*. The vertex is halfway between the focus and the directrix. As an example, let the focus be at point $(p, 0)$ and the directrix be the line $x = -p$.

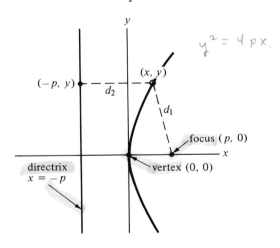

Here, the distances d_1 and d_2 are equal, where d_1 is the distance from any point (x, y) of the parabola to the focus $(p, 0)$ and d_2 is the distance from the point (x, y) to the directrix. The distance d_2 from a point to a line is measured on a perpendicular from (x, y) to the line $x = -p$, so the y coordinate of the point on $x = -p$ is the same as that of the point (x, y). Thus, d_2 is the distance between (x, y) and $(-p, y)$. The relationship $d_1 = d_2$ can be written as

$$\sqrt{(x - p)^2 + (y - 0)^2} = |x + p|$$

Squaring both sides yields

$$(x - p)^2 + y^2 = x^2 + 2px + p^2$$

or

$$x^2 - 2px + p^2 + y^2 = x^2 + 2px + p^2$$

which can be simplified to

$$y^2 = 4px$$

$\begin{cases} \text{equation of a } \textit{parabola} \\ \text{with focus } (p, 0) \text{ and} \\ \text{directrix } x = -p \end{cases}$

If the focus is $(0, p)$ and the directrix is $y = -p$, then the parabola appears as follows:

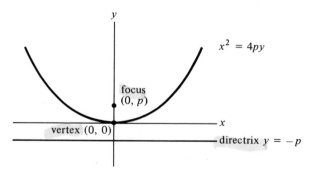

Its equation is

$$x^2 = 4py$$

$\begin{cases} \text{equation of } \textit{parabola} \\ \text{with focus } (0, p) \text{ and} \\ \text{directrix } y = -p \end{cases}$

Example 16. *Write the equation of the parabola with focus* $(0, -2)$ *and directrix* $y = 2$.

Here $p = -2$ in $x^2 = 4py$. So $x^2 = 4(-2)y$ or $x^2 = -8y$.

Example 17. *Determine the focus, directrix, and vertex of $y^2 = -2x$. Then sketch the graph of the parabola.*

The standard form is $y^2 = 4px$, and the given curve is $y^2 = -2x$. This means that $4p = -2$ or $p = -\frac{1}{2}$. So the focus is $(-\frac{1}{2}, 0)$ and the directrix is $x = \frac{1}{2}$. The vertex is at $(0, 0)$.

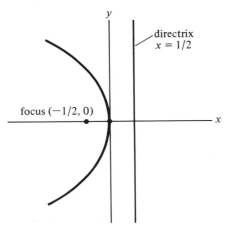

An *ellipse* is the set of points the sum of whose distances from two fixed points is constant. The two fixed points are called *foci*. The constant that represents the sum of the two distances will be called $2a$, for reasons that will become apparent later. If the foci are at $(c, 0)$ and $(-c, 0)$, then the ellipse appears as

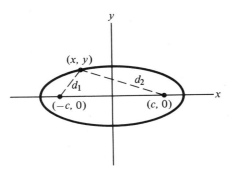

The sum of the distances $(d_1 + d_2)$ of any point (x, y) of the ellipse from $(-c, 0)$ and from $(c, 0)$ is equal to the constant $2a$.

$$d_1 + d_2 = 2a$$

Applying the distance formula to obtain values for d_1 and d_2 produces

$$\sqrt{(x + c)^2 + (y - 0)^2} + \sqrt{(x - c)^2 + (y - 0)^2} = 2a$$

Separate the radicals and then square both sides. This yields, in steps,

$$\sqrt{(x + c)^2 + y^2} = 2a - \sqrt{(x - c)^2 + y^2}$$

$$(x + c)^2 + y^2 = (2a)^2 + (x - c)^2 + y^2 - 4a\sqrt{(x - c)^2 + y^2}$$

$$x^2 + 2cx + c^2 + y^2 = 4a^2 + x^2 - 2cx + c^2 + y^2$$

$$- 4a\sqrt{(x - c)^2 + y^2}$$

$$4cx - 4a^2 = -4a\sqrt{(x - c)^2 + y^2}$$

$$cx - a^2 = -a\sqrt{(x - c)^2 + y^2}$$

$$a^2 - cx = a\sqrt{(x - c)^2 + y^2}$$

Now square both sides again.

$$(a^2 - cx)^2 = a^2[(x - c)^2 + y^2]$$

$$a^4 - 2a^2cx + c^2x^2 = a^2(x^2 - 2cx + c^2 + y^2)$$

$$a^4 - 2a^2cx + c^2x^2 = a^2x^2 - 2a^2cx + a^2c^2 + a^2y^2$$

$$a^4 + c^2x^2 = a^2x^2 + a^2c^2 + a^2y^2$$

This last form can be rewritten as

$$a^2x^2 - c^2x^2 + a^2y^2 = a^4 - a^2c^2$$

and factored as

$$(a^2 - c^2)x^2 + a^2y^2 = a^2(a^2 - c^2)$$

From the ellipse sketched earlier you can see that $2a$ (that is, $d_1 + d_2$) is greater than $2c$. Thus, $a^2 - c^2$ is a positive number. Let $b^2 = a^2 - c^2$. After making this substitution, the equation above simplifies to

$$b^2x^2 + a^2y^2 = a^2b^2$$

When both sides of this equation are divided by the nonzero number a^2b^2, the final result is

$$\boxed{\frac{x^2}{a^2} + \frac{y^2}{b^2} = 1}$$

$\left\{ \begin{array}{l} \text{equation of } \textit{ellipse} \text{ with} \\ \text{vertices at } (\pm a, 0) \text{ and} \\ (0, \pm b) \text{ and foci at } (\pm c, 0). \\ \text{Note } a > b. \end{array} \right.$

This is considered the standard form for the equation of an ellipse. The foci are at $(-c, 0)$ and $(c, 0)$, where c is determined from $b^2 = a^2 - c^2$. The vertices are on the x and y axes. If $x = 0$, $y = \pm b$. So $(0, b)$ and $(0, -b)$ are vertices. Of course, b and $-b$ are y intercepts, since they are the y values when x is zero. Similarly, if $y = 0$, then $x = \pm a$. So $(a, 0)$ and $(-a, 0)$ are vertices. The numbers a and $-a$ are x intercepts, since they are the x values when y is zero.

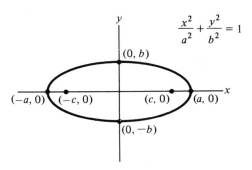

If the foci are at $(0, c)$ and $(0, -c)$, then the equation of the ellipse is

$$\frac{x^2}{b^2} + \frac{y^2}{a^2} = 1$$

(equation of *ellipse* with vertices at $(0, \pm a)$ and $(\pm b, 0)$ and foci at $(0, \pm c)$. Note $a > b$.

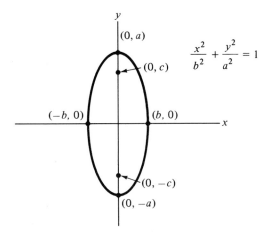

Example 18. *Sketch the ellipse* $\dfrac{x^2}{16} + \dfrac{y^2}{3} = 1$.

When y is zero, $x = \pm 4$. When x is zero, y is $\pm \sqrt{3}$. The vertices are then $(4, 0)$, $(-4, 0)$, $(0, \sqrt{3})$, and $(0, -\sqrt{3})$.

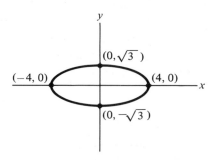

Although the foci are not needed here, they can be determined from $b^2 = a^2 - c^2$. Here $b^2 = 3$ and $a^2 = 16$, so $c^2 = 13$ or $c = \pm \sqrt{13}$. The foci are at $(\pm \sqrt{13}, 0)$.

Example 19. *Write the given equation of an ellipse in standard form:*

$$9x^2 + 4y^2 = 36.$$

Divide $9x^2 + 4y^2 = 36$ by 36. This will yield the $\ldots = 1$ form that is desired:

$$\frac{9x^2}{36} + \frac{4y^2}{36} = \frac{36}{36}$$

or

$$\frac{x^2}{4} + \frac{y^2}{9} = 1 \quad \checkmark$$

Example 20. *Determine the equation of the ellipse with vertices $(0, \pm 5)$ and foci $(0, \pm 2)$.*

We have $(0, \pm a) = (0, \pm 5)$ and $(0, \pm c) = (0, \pm 2)$. The standard form of the ellipse is

$$\frac{x^2}{b^2} + \frac{y^2}{a^2} = 1$$

The value of b can be computed using $b^2 = a^2 - c^2$.

$$b^2 = 25 - 4 = 21$$

The desired equation, then, is

$$\frac{x^2}{21} + \frac{y^2}{25} = 1 \quad \checkmark$$

The sketch:

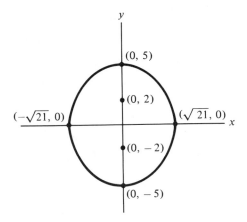

A *hyperbola* is the set of points the difference of whose distances from two fixed points is constant. The two fixed points are called *foci.* The constant representing the difference of the two distances will have absolute value $2a$. The distance $d_1 - d_2$ can be positive or negative. Thus, $d_1 - d_2 = \pm 2a$. The foci are at $(\pm c, 0)$.

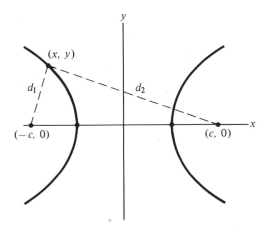

The relationship $d_1 - d_2 = \pm 2a$ leads to

$$\sqrt{(x + c)^2 + (y - 0)^2} - \sqrt{(x - c)^2 + (y - 0)^2} = \pm 2a$$

After a series of squarings and simplifications (as done with the ellipse relationship), a simpler, more useful form is derived. That form is

$$\frac{x^2}{a^2} - \frac{y^2}{b^2} = 1$$

$\begin{cases} \text{equation of } \textit{hyperbola} \\ \text{(where } b^2 = c^2 - a^2 \text{) with} \\ \text{vertices at } (\pm a, 0) \text{ and} \\ \text{foci at } (\pm c, 0) \end{cases}$

Note here that $c^2 > a^2$, unlike the ellipse. The vertices are $(\pm a, 0)$. The points $(0, \pm b)$ are not even on the hyperbola, for if $x = 0$, $y = \pm bi$.

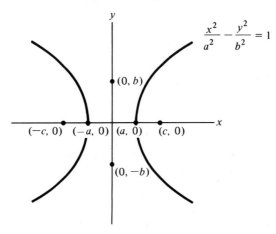

The hyperbola

$$\frac{y^2}{a^2} - \frac{x^2}{b^2} = 1$$

$\begin{cases} \text{equation of } \textit{hyperbola} \\ \text{(where } b^2 = c^2 - a^2 \text{) with} \\ \text{vertices at } (0, \pm a) \text{ and} \\ \text{foci at } (0, \pm c) \end{cases}$

is graphed as

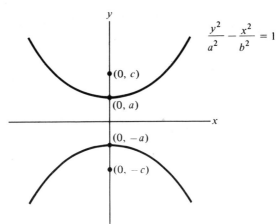

Example 21. *Sketch the graph of* $\frac{x^2}{16} - \frac{y^2}{9} = 1$.

The vertices are at $(\pm 4, 0)$.

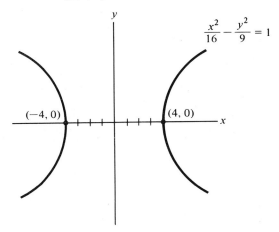

$b^2 = 9$, so $c^2 = a^2 + b^2 = 25$ yields $c = \pm 5$. The foci, although not needed here, are $(\pm 5, 0)$.

Example 22. *Sketch the graph of* $\frac{y^2}{25} - \frac{x^2}{7} = 1$.

The vertices are at $(0, \pm 5)$.

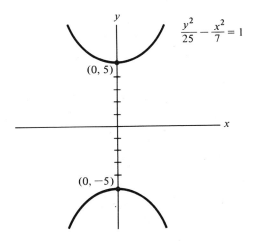

Here $b^2 = 7$, so $c^2 = a^2 + b^2 = 32$ yields $c = \pm\sqrt{32}$ or $\pm 4\sqrt{2}$. The foci are $(0, \pm 4\sqrt{2})$.

EXERCISE SET C

*1. The *unit circle* is the circle with center at the origin and radius 1. Write the equation of the unit circle.

*2. Write the equation of the circle with the given point as center and the given radius.

(a) $(3, 12)$, $r = 4$ (b) $(1, 0)$, $r = 2$

(c) $(-2, -7)$, $r = 10$ (d) $(0, -9)$, $r = 1$

*3. Determine the center and radius of each circle.

(a) $(x - 4)^2 + (y - 5)^2 = 49$ (b) $(x - 9)^2 + (y + 8)^2 = 64$

(c) $(x + 1)^2 + (y + 2)^2 = 5$ (d) $x^2 + (y - 1)^2 = 12$

4. Determine the center and radius of each circle.

  *(a) $x^2 + y^2 + 4x + 6y + 12 = 0$

  *(b) $x^2 + y^2 + 12x - 10y + 25 = 0$

  *(c) $x^2 + y^2 - 6x - 2y - 54 = 0$

  (d) $x^2 + y^2 - 14x - 51 = 0$

  (e) $x^2 + y^2 + 4x - 6y + 8 = 0$

5. Find the equation of the circle which has its center at $(4, -3)$ and passes through $(2, 0)$.

6. Determine the vertex, focus, and directrix of each parabola. Then sketch its graph.

  *(a) $y^2 = 8x$ (b) $y^2 = -8x$ (c) $x^2 = 12y$

  *(d) $x^2 = -12y$ *(e) $y^2 = 10x$ (f) $y^2 = x$

  *(g) $x^2 - 2y = 0$ (h) $3x^2 - 4y = 0$ *(i) $2x + 3y^2 = 0$

  (j) $y = \pm 2\sqrt{x}$

7. Write the equation of the parabola with given focus and directrix.

  *(a) focus $(5, 0)$, directrix $x = -5$ (b) focus $(-5, 0)$, directrix $x = 5$

  *(c) focus $(0, \frac{1}{2})$, directrix $y = -\frac{1}{2}$ (d) focus $(0, -2)$, directrix $y = 2$

  *(e) focus $(1, 0)$, directrix $x + 1 = 0$

  (f) focus $(0, -3)$, directrix $y - 3 = 0$

8. Determine the vertices and foci, and sketch each ellipse or hyperbola.

  *(a) $\dfrac{x^2}{9} + \dfrac{y^2}{1} = 1$ (b) $\dfrac{x^2}{25} + \dfrac{y^2}{4} = 1$ *(c) $\dfrac{x^2}{16} + \dfrac{y^2}{10} = 1$

     *Answers to starred problems are to be found at the back of the book.

(d) $\dfrac{x^2}{16} - \dfrac{y^2}{10} = 1$ ⋆(e) $\dfrac{x^2}{9} - \dfrac{y^2}{4} = 1$ (f) $\dfrac{y^2}{16} - \dfrac{x^2}{9} = 1$

⋆(g) $9x^2 + 4y^2 = 36$ ⋆(h) $9y^2 - 4x^2 = 36$ (i) $25x^2 + 4y^2 = 400$

(j) $25x^2 - 4y^2 = 400$

6.5 TRANSLATION

The equations of parabolas, ellipses, and hyperbolas developed in the last section were for *central conics*, that is, conics symmetric with respect to the origin. On the other hand, circles other than those centered at the origin were included. The equation of the circle was considered for all centers (h, k). Specifically, circles centered at the origin (and therefore symmetric with respect to the origin) can be written in the form

$$x^2 + y^2 = r^2$$

Circles centered at (h, k) can be written as

$$(x - h)^2 + (y - k)^2 = r^2$$

Such circles can be thought of as translated from center at $(0, 0)$ to center at (h, k).

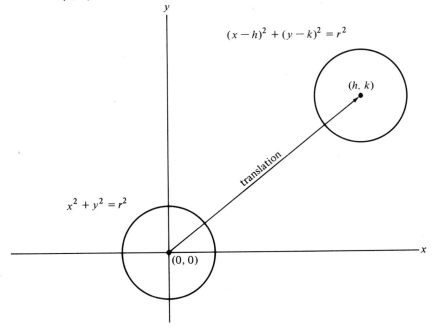

This translation can also be viewed as the creation of a new coordinate system of points (x', y'), in which (h, k) is the origin. That is,

$$x' = x - h \qquad (or\ x = x' + h)$$
$$y' = y - k \qquad (or\ y = y' + k)$$

so that

$$(x - h)^2 + (y - k)^2 = r^2$$

becomes

$$(x')^2 + (y')^2 = r^2$$

When similar translations are applied to the other central conics (parabolas, ellipses, and hyperbolas), the resulting conics are:

1. *Parabola* with vertex at (h, k), focus at $(p + h, k)$, and directrix $x = -p + h$.

$$\boxed{(y - k)^2 = 4p(x - h)}$$

2. *Parabola* with vertex at (h, k), focus at $(h, p + k)$, and directrix $y = -p + k$.

$$\boxed{(x - h)^2 = 4p(y - k)}$$

3. *Ellipse* with vertices at $(h \pm a, k)$ and $(h, k \pm b)$ and foci at $(h \pm c, k)$.

$$\boxed{\frac{(x - h)^2}{a^2} + \frac{(y - k)^2}{b^2} = 1} \qquad (c^2 = a^2 - b^2)$$

4. *Ellipse* with vertices at $(h \pm b, k)$ and $(h, k \pm a)$ and foci at $(h, k \pm c)$.

$$\boxed{\frac{(x - h)^2}{b^2} + \frac{(y - k)^2}{a^2} = 1} \qquad (c^2 = a^2 - b^2)$$

5. *Hyperbola* with foci at $(h \pm c, k)$ and vertices $(h \pm a, k)$.

$$\boxed{\frac{(x - h)^2}{a^2} - \frac{(y - k)^2}{b^2} = 1} \qquad (c^2 = a^2 + b^2)$$

6. *Hyperbola* with foci at $(h, k \pm c)$ and vertices at $(h, k \pm a)$.

$$\boxed{\frac{(y - k)^2}{a^2} - \frac{(x - h)^2}{b^2} = 1} \qquad (c^2 = a^2 + b^2)$$

Example 23. *Sketch the graph of the conic.*

$$\frac{(x - 3)^2}{25} + \frac{(y + 1)^2}{4} = 1$$

This is an ellipse centered about $(3, -1)$ rather than $(0, 0)$. Since a is 5 and b is 2, the vertices are at $(3 \pm 5, -1)$ and $(3, -1 \pm 2)$. The coordinates of the vertices, when simplified, are $(8, -1)$, $(-2, -1)$, $(3, 1)$, and $(3, -3)$.

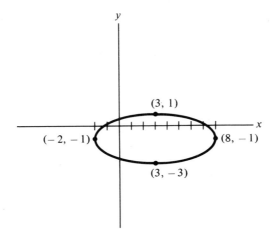

Since $c^2 = a^2 - b^2$, $c^2 = 25 - 4 = 21$. So $c = \pm\sqrt{21}$. This means that the foci are at $(3 \pm \sqrt{21}, -1)$.

Example 24. *Identify and sketch the conic:*

$$4x^2 - 9y^2 + 8x + 90y - 257 = 0.$$

The $4x^2 - 9y^2$ portion suggests that this curve is a hyperbola. Accordingly, we should be able to get it into the standard form,

$$\frac{(x - h)^2}{a^2} - \frac{(y - k)^2}{b^2} = 1$$

from which graphing will be straightforward. The standard form contains the completed squares $(x - h)^2$ and $(y - k)^2$. In view of this, terms should be grouped for completing the squares.

$$4x^2 + 8x - 9y^2 + 90y = 257$$

Factor 4 out of $4x^2$ and $8x$. Factor -9 out of $-9y^2$ and $90y$.

$$4(x^2 + 2x) - 9(y^2 - 10y) = 257$$

The desired square for $x^2 + 2x$ is $(x + 1)^2$, which is 1 more than $x^2 + 2x$. Also, this quantity is multiplied by 4. Thus, $4(x + 1)^2$ is 4 more than $4(x^2 + 2x)$. So add 4 to the right side of the equation to complete that square.

$$4(x + 1)^2 - 9(y^2 - 10y) = 257 + 4$$

The desired square for $y^2 - 10y$ is $(y - 5)^2$, which is 25 more than $y^2 - 10y$. Also, this quantity is multiplied by -9. Thus, $-9(y - 5)^2$ is -225 more than $-9(y^2 - 10y)$. So add -225 to the right side of the equation to complete the square. The result is

$$4(x + 1)^2 - 9(y - 5)^2 = 257 + 4 - 225$$

or

$$4(x + 1)^2 - 9(y - 5)^2 = 36$$

Divide both sides of the equation by 36 to get the standard form.

$$\frac{(x + 1)^2}{9} - \frac{(y - 5)^2}{4} = 1 \qquad \begin{cases} a \text{ is } 3 \\ b \text{ is } 2 \end{cases}$$

The hyperbola is centered about $(-1, 5)$ rather than $(0, 0)$. The vertices are at $(-1 \pm 3, 5)$, that is, $(2, 5)$ and $(-4, 5)$.

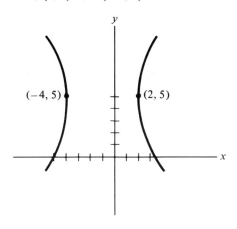

Example 25. *Sketch $x^2 + 6x - 8y + 17 = 0$.*

This is a parabola. It can be manipulated into the form $(x - h)^2 = 4p(y - k)$. First, collect the x^2 and x terms on the left side and complete the square.

$$x^2 + 6x = 8y - 17$$

or

$$(x + 3)^2 = 8y - 17 + 9$$
$$(x + 3)^2 = 8y - 8$$

Now factor 8, the coefficient of y, from $8y - 8$. The result is the desired form.

$$(x + 3)^2 = 8(y - 1) \qquad (4p = 8,\ p = 2)$$

The vertex is at $(-3, 1)$. The focus is at $(h, k + p) = (-3, 1 + 2) = (-3, 3)$. The directrix is $y = -p + k$ or $y = -1$.

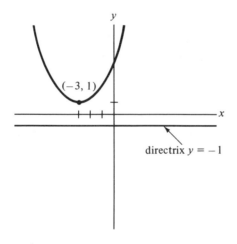

EXERCISE SET D

1. Write each relation in standard form and sketch its graph. For parabolas, find vertex and directrix. For ellipses and hyperbolas, find the vertices and foci.

   ★(a) $4(x - 2)^2 + 9(y + 1)^2 = 36$

   (b) $(x + 5)^2 + 16(y + 3)^2 = 16$

   ★(c) $25(x - 3)^2 - 4y^2 = 100$

   (d) $(x - 5)^2 - 3(y + 2) = 0$

★Answers to starred problems are to be found at the back of the book.

$\star$(e) $y^2 - 7x = 21$

$\star$(f) $4x^2 + y^2 + 16x - 2y + 13 = 0$

$\star$(g) $9x^2 - 16y^2 - 160y - 544 = 0$

$\star$(h) $y^2 + 6y - 9x - 9 = 0$

(i) $5x^2 + 3y^2 - 70x + 12y + 242 = 0$

(j) $9y^2 - 2x^2 - 18y - 36x - 157 = 0$

(k) $2(x - 7)^2 = 28 - 7(y + 4)^2$

(l) $3x^2 - 6x - 6y + 1 = 0$

$\star$**2.** The *general form of conics* is

$$Ax^2 + Bxy + Cy^2 + Dx + Ey + F = 0$$

If $B^2 - 4AC = 0$, the conic is a parabola. If $B^2 - 4AC < 0$, the conic is an ellipse. If $B^2 - 4AC > 0$, the conic is a hyperbola. Test the conics to determine the nature of each.

(a) $3x^2 + 2xy + y^2 + 5x - 2y + 7 = 0$

(b) $x^2 - 5xy + y^2 + 3y - 9 = 0$

(c) $x^2 + 2xy + y^2 - x + 5y = 0$

(d) $xy + 7x + 5y - 18 = 0$

(e) $9x^2 + 7y^2 - 8x + 2y - 3 = 0$

chapter seven

---

systems of equations

7.1 INTRODUCTION

Many problems in mathematics involve finding numbers which satisfy simultaneously a number of linear equations. In this chapter, we explain techniques for solving such systems and others.

The graph of any linear function is a straight line. When two such functions are graphed together on the same axes, one of three things must occur. The lines might meet in one point, the lines might be parallel, or, the "two" lines may indeed be the same line. This interpretation is based on graphs. Two linear relations in x and y can be manipulated algebraically in order to find the values of x and y of the point (x, y) where the two lines meet. In other words, you can determine the value of x and the value of y which together satisfy both equations. And once again, one of three things will occur when you attempt to solve the system algebraically. If a solution is obtained, the equations are called *consistent*. If no solution is obtained, and instead results such as $4 = 0$ or $5 = 8$ are produced, the equations are called *inconsistent*. If no solution is obtainable, but results such as $0 = 0$ or $7 = 7$ are produced, then the equations are called *dependent*.

An approximation to the solution of the system

$$\begin{cases} y = 3x + 1 \\ 2x + 3y = 12 \end{cases}$$

can be obtained by graphing.

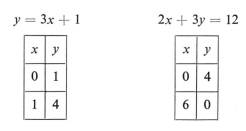

$y = 3x + 1$

x	y
0	1
1	4

$2x + 3y = 12$

x	y
0	4
6	0

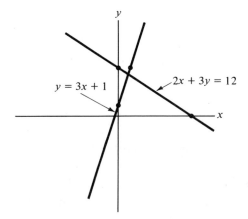

The lines are consistent; they appear to meet at about $(1, 3\frac{1}{2})$. The exact point of intersection can be determined by algebraic techniques to be $(\frac{9}{11}, \frac{38}{11})$. This will be done here by a method called *substitution*.

The first equation is $y = 3x + 1$, which states that y is the same as $3x + 1$. If the expression $3x + 1$ is substituted for y in the other equation, then that equation will contain only one variable, x. Thus,

$$2x + 3(y) = 12$$

becomes

$$2x + 3(3x + 1) = 12$$

when $3x + 1$ is substituted for y. The result is an equation in x alone that is easily solved.

$$2x + 3(3x + 1) = 12$$
$$2x + 9x + 3 = 12$$
$$11x + 3 = 12$$
$$11x = 9$$
$$x = \frac{9}{11} \quad \checkmark$$

The value of y can now be found by using the substitution equation $y = 3x + 1$. Since $x = \frac{9}{11}$,

$$y = 3\left(\frac{9}{11}\right) + 1$$

$$= \frac{27}{11} + 1$$

$$= \frac{27}{11} + \frac{11}{11}$$

$$= \frac{38}{11} \quad \checkmark$$

The solution is $x = \frac{9}{11}$, $y = \frac{38}{11}$. This process of substitution is natural to use when one of the two linear equations is written as $y = \cdots$ or as $x = \cdots$. Otherwise this technique is usually a bad choice.

The system

$$\begin{cases} 3x + 2y = 13 \\ 5x - 4y = 7 \end{cases}$$

can be solved by substitution if you transform one equation into $y = \cdots$ or $x = \cdots$ for substitution into the other equation. Such a process lacks appeal. Instead, multiply both sides of the top equation by 2. The modified system is

$$\begin{cases} 6x + 4y = 26 \\ 5x - 4y = 7 \end{cases}$$

Now add the two equations term for term. That is, add $5x - 4y$ to $6x + 4y$ and add 7 to 26. This is an example of adding equal quantities to equal quantities to produce two equal quantities. The resulting new equation is

$$11x + 0y = 33$$

or simply

$$11x = 33$$

from which

$$x = 3 \quad \checkmark$$

Return to either of the original equations to determine y. Let's use the first one, $3x + 2y = 13$. Since x is 3, the equation

$$3(x) + 2y = 13$$

becomes

$$3(3) + 2y = 13$$
$$9 + 2y = 13$$
$$2y = 4$$
$$y = 2 \quad \checkmark$$

The solution is $x = 3$, $y = 2$.

Return to the original equations and note that the top equation was multiplied by 2 in order to make the coefficients of y equal in magnitude (4) but opposite in sign (one $+$, one $-$). In this way they add to zero, thus eliminating one of the variables (y).

The technique used above, in which one of the variables is eliminated by addition, is called *elimination* or *addition*. In the next example, we consider a system in which the elimination process is a little more complicated.

Example 1. *Solve the system by elimination.*

$$\begin{cases} 3x + 7y = -5 \\ 2x + 9y = \quad 1 \end{cases}$$

If the top equation is multiplied by -2 and the bottom equation is multiplied by 3, then the coefficients of x will "agree" at -6 and 6 and thus add to zero.

$$\begin{cases} 3x + 7y = -5 \xrightarrow{-2} -6x - 14y = 10 \\ 2x + 9y = \quad 1 \xrightarrow{\ 3\ } \quad 6x + 27y = \quad 3 \end{cases}$$
$$\overline{\hspace{3cm} 13y = 13}$$
$$y = 1 \quad \checkmark$$

The value of x can now be determined by using one of the original equations, say $2x + 9y = 1$, with $y = 1$.

$$2x + 9(y) = 1$$
$$2x + 9(1) = 1$$
$$2x + 9 = 1$$
$$2x = -8$$
$$x = -4 \quad \checkmark$$

The solution is $x = -4$, $y = 1$.

Let's check the solution to be sure that it does indeed satisfy both equations. If it does not, then we had better look for the error or redo the work.

First equation: $\qquad\qquad 3x + 7y = -5$

Let $x = -4$, $y = 1$. $\quad 3(-4) + 7(1) \overset{?}{=} -5$

$$-12 + 7 \overset{?}{=} -5$$
$$-5 = -5 \quad \checkmark \qquad \text{(It checks.)}$$

Second equation: $\qquad\qquad 2x + 9y = 1$

Let $x = -4$, $y = 1$. $\quad 2(-4) + 9(1) \overset{?}{=} 1$

$$-8 + 9 \overset{?}{=} 1$$
$$1 = 1 \quad \checkmark \qquad \text{(It checks.)}$$

A system of three equations in three unknowns can be solved by reducing the system to two equations in two unknowns. We shall now use elimination to do that.

Example 2. *Solve the following system.*

$$\begin{cases} 5x + 2y + z = 5 \\ 2x + 3y - z = -6 \\ 7x + 4y + 2z = 4 \end{cases}$$

If the first and second equations are added, z will be eliminated.

$$\begin{array}{l} 5x + 2y + z = 5 \\ 2x + 3y - z = -6 \\ \hline 7x + 5y \quad\;\; = -1 \quad \checkmark \end{array}$$

Returning to the original equations, note that if twice the second equation is added to the third equation, z will be eliminated.

$$2x + 3y - z = -6 \xrightarrow{2} 4x + 6y - 2z = -12$$
$$7x + 4y + 2z = 4 \longrightarrow 7x + 4y + 2z = 4$$
$$\overline{11x + 10y = -8} \checkmark$$

We now have two equations in the same two unknowns, x and y: $7x + 5y = -1$ and $11x + 10y = -8$. These can be combined as one system and easily solved for x and y. Specifically, multiply $7x + 5y = -1$ by -2 in order to eliminate y.

$$\begin{cases} 7x + 5y = -1 \\ 11x + 10y = -8 \end{cases} \xrightarrow{-2} \begin{aligned} -14x - 10y &= 2 \\ 11x + 10y &= -8 \\ \hline -3x &= -6 \\ x &= 2 \checkmark \end{aligned}$$

Now y can be determined by using 2 for x in the equation $7x + 5y = -1$.

$$7(2) + 5y = -1$$
$$14 + 5y = -1$$
$$5y = -15$$
$$y = -3 \checkmark$$

Finally, return to the original system, select any equation, and determine z by substituting 2 for x and -3 for y. The first equation, $5x + 2y + z = 5$, appears to be the easiest to use because the coefficient of the unknown z is simply 1.

$$5x + 2y + z = 5$$
$$5(2) + 2(-3) + z = 5$$
$$10 - 6 + z = 5$$
$$4 + z = 5$$
$$z = 1 \checkmark$$

The solution is

$$x = 2$$
$$y = -3$$
$$z = 1$$

Each linear equation in three variables represents a plane in three dimensions. The solution to the system, then, can be visualized as the point where the three planes meet. This is an (x, y, z) point in three dimensions. For the example above, the planes meet at $(2, -3, 1)$.

EXERCISE SET A

1. Solve each pair of simultaneous linear equations by graphing the lines and approximating the point of intersection, if any.

$\star$(a) $\begin{cases} y = 2x + 1 \\ y = -x + 3 \end{cases}$ $\star$(b) $\begin{cases} y = 3x - 2 \\ y = x \end{cases}$

$\star$(c) $\begin{cases} y = 4x - 1 \\ y = -4x + 3 \end{cases}$ $\star$(d) $\begin{cases} y = 3 - x \\ y = 2x + 2 \end{cases}$

(e) $\begin{cases} x + y - 5 = 0 \\ x - y = 0 \end{cases}$ (f) $\begin{cases} 2x - y = 3 \\ x + 3y = 6 \end{cases}$

(g) $\begin{cases} 3x + 2y = 6 \\ 3y - 2x = 12 \end{cases}$ (h) $\begin{cases} 3y - 4x = 6 \\ 2y + 5x = 10 \end{cases}$

$\star$(i) $\begin{cases} y = x + 5 \\ x = -3 \end{cases}$ (j) $\begin{cases} x = 3 \\ y = -5 \end{cases}$

2. Solve each pair of simultaneous linear equations by using substitution. Then check the solution in both equations.

$\star$(a) $\begin{cases} 2x + y = 11 \\ y = x + 2 \end{cases}$ $\star$(b) $\begin{cases} x + 4y = 8 \\ x = 3y + 1 \end{cases}$

(c) $\begin{cases} 2x + 3y = 15 \\ y = x \end{cases}$ (d) $\begin{cases} 5x - 8y = 9 \\ x = 2y + 1 \end{cases}$

$\star$(e) $\begin{cases} 5x - 7y = 16 \\ x = 6 \end{cases}$ $\star$(f) $\begin{cases} 5x - 2y = 39 \\ x = 1 - 3y \end{cases}$

(g) $\begin{cases} y = 5x + 1 \\ 4x + 3y = 10 \end{cases}$ (h) $\begin{cases} 4x - 3y = 9 \\ y = 7 - 2x \end{cases}$

$\star$(i) $\begin{cases} 2x - y + 6 = 0 \\ 3x - 5y = 5 \end{cases}$ (j) $\begin{cases} x + 2y = 1 \\ 5x - 3y = 13 \end{cases}$

3. Solve each pair of simultaneous linear equations by using elimination. Then check the solution in both equations.

$\star$(a) $\begin{cases} x + y = 6 \\ x - y = 4 \end{cases}$ (b) $\begin{cases} x - 3y = 1 \\ 2x + 3y = 20 \end{cases}$

$\star$(c) $\begin{cases} 2x + y = 7 \\ 5x - 3y = 23 \end{cases}$ $\star$(d) $\begin{cases} 3x + 7y = 7 \\ x + 5y = -3 \end{cases}$

$\star$Answers to starred problems are to be found at the back of the book.

(e) $\begin{cases} 2x + 5y = -1 \\ 6x + 7y = 5 \end{cases}$

*(f) $\begin{cases} 3x + 2y = 6 \\ -2x - 5y = 7 \end{cases}$

*(g) $\begin{cases} 5x + 2y = 7 \\ 7x + 3y = 9 \end{cases}$

(h) $\begin{cases} 3x + 5y = 11 \\ 4x + 2y = -4 \end{cases}$

(i) $\begin{cases} 7x + 3y = 1 \\ 5x + 7y = 8 \end{cases}$

(j) $\begin{cases} 8x - 4y = 7 \\ 5x - 9y = 6 \end{cases}$

4. Solve each system of equations and check your solution.

*(a) $\begin{cases} x + y + z = 6 \\ 2x + 3y - z = 5 \\ 3x + 2y + z = 10 \end{cases}$

(b) $\begin{cases} 5x - y + 2z = 8 \\ 2x + y + 5z = 6 \\ 3x - y + 3z = 7 \end{cases}$

*(c) $\begin{cases} 5x + 3y + z = 7 \\ x + 4y - 2z = 8 \\ 3x + 7y - 4z = 18 \end{cases}$

(d) $\begin{cases} x - 4y + 3z = 1 \\ -2x + 3y + 2z = -7 \\ 3x + 2y + 5z = 17 \end{cases}$

*(e) $\begin{cases} 7x + 3y + 2z = 9 \\ 3x - y + 3z = 17 \\ 5x + 4y + 4z = 13 \end{cases}$

(f) $\begin{cases} 2x - 3y + 4z = -3 \\ 4x + 5y - 5z = 31 \\ 6x + 2y + 3z = 20 \end{cases}$

*(g) $\begin{cases} 2r + 3s + 2t = 3 \\ 3r + 5s + 7t = 19 \\ 6r + 4s + 3t = 12 \end{cases}$

(h) $\begin{cases} 5m + 2n - 3p = -8 \\ 3m - 3n + 5p = 13 \\ -4m + 2n - 2p = -6 \end{cases}$

*(i) $\begin{cases} 5x + 2y + 3z = 1 \\ 7x - 3y + 7z = 16 \\ x + 5y - 5z = -8 \end{cases}$

7.2 DETERMINANTS

In the last section, the elimination and substitution methods were used to solve systems of linear equations. Now the elimination method is applied to the general system of two equations in two unknowns:

$$\begin{cases} ax + by = k_1 \\ cx + dy = k_2 \end{cases}$$

where a, b, c, d, k_1, and k_2 are constants.

Multiply the first equation by $-c$ and the second equation by a. This will make the coefficients of x equal to $-ac$ and ac. They will add to zero, thus eliminating x.

$$\begin{array}{lll} ax + by = k_1 \xrightarrow{-c} -acx - cby & = -ck_1 \\ cx + dy = k_2 \xrightarrow{a} acx + ady & = ak_2 \\ \hline \phantom{cx + dy = k_2 \xrightarrow{a}}0 + ady - cby = ak_2 - ck_1 \end{array}$$

Thus far

$$ady - cby = ak_2 - ck_1$$

Factor out y on the left side.

$$(ad - cb)y = ak_2 - ck_1$$

Next, divide both sides of the equation by the coefficient of y, namely, $ad - cb$. The result is

$$y = \frac{ak_2 - ck_1}{ad - cb}$$

Upon close examination this formula for y reveals the following. The denominator, $ad - bc$, is obtained from a times d minus c times b.

$$\begin{align} ax + by &= k_1 \\ cx + dy &= k_2 \end{align}$$

To aid in visualizing and remembering this pattern, we define a *determinant* called

$$\begin{vmatrix} a & b \\ c & d \end{vmatrix}$$

as

$$\begin{vmatrix} a & b \\ c & d \end{vmatrix} = a \cdot d - c \cdot b$$

The numerator, $ak_2 - ck_1$, can now be written as a determinant.

$$\begin{vmatrix} a & k_1 \\ c & k_2 \end{vmatrix} = ak_2 - ck_1$$

Putting these two determinants together gives

$$y = \frac{\begin{vmatrix} a & k_1 \\ c & k_2 \end{vmatrix}}{\begin{vmatrix} a & b \\ c & d \end{vmatrix}}$$

if the value of the denominator determinant is not zero.

Observe the following:

1. The denominator determinant contains the coefficients of x and y from the original equations.
2. The numerator determinant contains the coefficients of x. The constants k_1 and k_2 are used instead of the coefficients of y.

If the original equations are solved for x, the result is

$$x = \frac{k_1 d - k_2 b}{ad - cb}$$

or

$$x = \frac{\begin{vmatrix} k_1 & b \\ k_2 & d \end{vmatrix}}{\begin{vmatrix} a & b \\ c & d \end{vmatrix}}$$

The denominator is the same determinant that was used for the denominator of y. The numerator determinant contains the y coefficients, but the constants k_1 and k_2 are used in place of the x coefficients. In summary, if

$$\begin{cases} ax + by = k_1 \\ cx + dy = k_2 \end{cases}$$

then

$$x = \frac{\begin{vmatrix} k_1 & b \\ k_2 & d \end{vmatrix}}{\begin{vmatrix} a & b \\ c & d \end{vmatrix}} \qquad y = \frac{\begin{vmatrix} a & k_1 \\ c & k_2 \end{vmatrix}}{\begin{vmatrix} a & b \\ c & d \end{vmatrix}}$$

This method of using determinants to solve systems of linear equations is called *Cramer's rule*. It was published about 1750 by the Swiss mathematician Gabriel Cramer.

Example 3. *Solve the system of linear equations by using Cramer's rule.*

The system

$$\begin{cases} 3x - 2y = 7 \\ 4x + 5y = 1 \end{cases}$$

is solved as follows:

$$x = \frac{\begin{vmatrix} 7 & -2 \\ 1 & 5 \end{vmatrix}}{\begin{vmatrix} 3 & -2 \\ 4 & 5 \end{vmatrix}} = \frac{(7)(5) - (1)(-2)}{(3)(5) - (4)(-2)} = \frac{37}{23} \checkmark$$

$$y = \frac{\begin{vmatrix} 3 & 7 \\ 4 & 1 \end{vmatrix}}{\begin{vmatrix} 3 & -2 \\ 4 & 5 \end{vmatrix}} = \frac{(3)(1) - (4)(7)}{23} = \frac{-25}{23} = -\frac{25}{23} \checkmark$$

The general system of three linear equations in three variables, such as

$$\begin{cases} ax + by + cz = k_1 \\ dx + ey + fz = k_2 \\ gx + hy + jz = k_3 \end{cases}$$

can be solved by the addition method (elimination) to produce

$$x = \frac{(k_1 ej + bfk_3 + ck_2 h) - (k_3 ec + hfk_1 + jk_2 b)}{(aej + bfg + cdh) - (gec + hfa + jdb)}$$

$$y = \frac{(ak_2 j + k_1 fg + ck_3 d) - (gk_2 c + k_3 fa + jk_1 d)}{(aej + bfg + cdh) - (gec + hfa + jdb)}$$

$$z = \frac{(aek_3 + bk_2 g + k_1 hd) - (gek_1 + hk_2 a + k_3 bd)}{(aej + bfg + cdh) - (gec + hfa + jdb)}$$

Note that the denominator is the same for all three variables x, y, and z. Only the coefficients of x, y, and z are used (that is, only a, b, c, d, e, f, g, h, and j). We can write the denominator as the determinant

$$\begin{vmatrix} a & b & c \\ d & e & f \\ g & h & j \end{vmatrix}$$

to compare with the determinant used for two equations involving two variables. We now seek a mechanical way of computing

$$\begin{vmatrix} a & b & c \\ d & e & f \\ g & h & j \end{vmatrix} = (aej + bfg + cdh) - (gec + hfa + jdb)$$

Copy over the first two columns of the determinant, placing the two columns to the right of the determinant, as

$$\begin{vmatrix} a & b & c \\ d & e & f \\ g & h & j \end{vmatrix} \begin{matrix} a & b \\ d & e \\ g & h \end{matrix}$$

Now observe

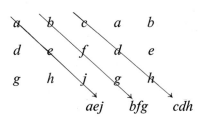

and

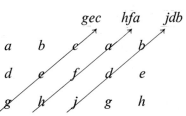

Add the first three products ($aej + bfg + cdh$) and *subtract* from that sum the sum of the other three products ($gec + hfa + jdb$).

Example 4. *Compute the value of the determinant*

$$\begin{vmatrix} 1 & 2 & 4 \\ 3 & 7 & 5 \\ 9 & 6 & 8 \end{vmatrix}$$

Rewrite the determinant as

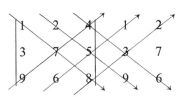

Its value is then

$$(1 \cdot 7 \cdot 8 + 2 \cdot 5 \cdot 9 + 4 \cdot 3 \cdot 6) - (9 \cdot 7 \cdot 4 + 6 \cdot 5 \cdot 1 + 8 \cdot 3 \cdot 2)$$

or

$$(56 + 90 + 72) - (252 + 30 + 48) = -112 \quad \checkmark$$

The value of the determinant is -112.

Let us return now to the expressions for x, y, and z. It has already been shown that the denominator for each of them is

$$\begin{vmatrix} a & b & c \\ d & e & f \\ g & h & j \end{vmatrix}$$

Without too much trouble, the numerators can be written as determinants. They are shown next along with the previously determined denominator determinant.

$$x = \frac{\begin{vmatrix} k_1 & b & c \\ k_2 & e & f \\ k_3 & h & j \end{vmatrix}}{\begin{vmatrix} a & b & c \\ d & e & f \\ g & h & j \end{vmatrix}}$$

$\left\{\begin{array}{l}\text{The column of } x \text{ coefficients} \\ \text{has been replaced by the} \\ \text{column of constants } k_1, k_2, \\ k_3.\end{array}\right.$

$$y = \frac{\begin{vmatrix} a & k_1 & c \\ d & k_2 & f \\ g & k_3 & j \end{vmatrix}}{\begin{vmatrix} a & b & c \\ d & e & f \\ g & h & j \end{vmatrix}}$$

$\left\{\begin{array}{l}\text{The column of } y \text{ coefficients} \\ \text{has been replaced by the} \\ \text{column of constants } k_1, k_2, \\ k_3,\end{array}\right.$

$$z = \frac{\begin{vmatrix} a & b & k_1 \\ d & e & k_2 \\ g & h & k_3 \\ a & b & c \\ d & e & f \\ g & h & j \end{vmatrix}}{}$$

$\left\{\begin{array}{l}\text{The column of } z \text{ coefficients} \\ \text{has been replaced by the} \\ \text{column of constants } k_1, k_2, \\ k_3.\end{array}\right.$

Example 5. *Solve the following system of linear equations by using determinants (Cramer's rule).*

$$\begin{cases} 2x + y - 5z = 7 \\ 8x - 3y + z = 4 \\ 7x + 9y + 6z = 0 \end{cases}$$

The solution is as follows:

$$x = \frac{\begin{vmatrix} 7 & 1 & -5 \\ 4 & -3 & 1 \\ 0 & 9 & 6 \end{vmatrix}}{\begin{vmatrix} 2 & 1 & -5 \\ 8 & -3 & 1 \\ 7 & 9 & 6 \end{vmatrix}} = \frac{(-306) - (87)}{(-389) - (171)} = \frac{-393}{-560} = \frac{393}{560} \checkmark$$

$$y = \frac{\begin{vmatrix} 2 & 7 & -5 \\ 8 & 4 & 1 \\ 7 & 0 & 6 \end{vmatrix}}{\begin{vmatrix} 2 & 1 & -5 \\ 8 & -3 & 1 \\ 7 & 9 & 6 \end{vmatrix}} = \frac{(97) - (196)}{-560} = \frac{-99}{-560} = \frac{99}{560} \checkmark$$

$$z = \frac{\begin{vmatrix} 2 & 1 & 7 \\ 8 & -3 & 4 \\ 7 & 9 & 0 \end{vmatrix}}{\begin{vmatrix} 2 & 1 & -5 \\ 8 & -3 & 1 \\ 7 & 9 & 6 \end{vmatrix}} = \frac{(532) - (-75)}{-560} = \frac{607}{-560} = -\frac{607}{560} \checkmark$$

If you think this was a lot of work, imagine doing it by the addition method with such large fractions! One nice feature of Cramer's rule is that fractions are essentially eliminated from the computation.

EXERCISE SET B

*1. Compute the value of each determinant.

(a) $\begin{vmatrix} 4 & 1 \\ 2 & 3 \end{vmatrix}$

(b) $\begin{vmatrix} 5 & 8 \\ 0 & 0 \end{vmatrix}$

(c) $\begin{vmatrix} 4 & 7 \\ 0 & 3 \end{vmatrix}$

(d) $\begin{vmatrix} 0 & 9 \\ 0 & 8 \end{vmatrix}$

(e) $\begin{vmatrix} 5 & 1 \\ 4 & -3 \end{vmatrix}$

(f) $\begin{vmatrix} 15 & 6 \\ -2 & 4 \end{vmatrix}$

(g) $\begin{vmatrix} 5 & 5 \\ 5 & 5 \end{vmatrix}$

(h) $\begin{vmatrix} 2 & 4 \\ 6 & 12 \end{vmatrix}$

(i) $\begin{vmatrix} 1 & 2 \\ -2 & -4 \end{vmatrix}$

(j) $\begin{vmatrix} \sin x & -\cos x \\ \cos x & \sin x \end{vmatrix}$

2. Solve each system of linear equations by using Cramer's rule.

*(a) $\begin{cases} x + y = 7 \\ -2x + 5y = 14 \end{cases}$

*(b) $\begin{cases} x + 2y = 0 \\ 3x - y = 0 \end{cases}$

*(c) $\begin{cases} 2x - 3y = 11 \\ x - y = 3 \end{cases}$

(d) $\begin{cases} 2x + 2y = 12 \\ x + 3y = 14 \end{cases}$

*Answers to starred problems are to be found at the back of the book.

*(e) $\begin{cases} 4p + q = 1 \\ 4p - 2q = 6 \end{cases}$ (f) $\begin{cases} 5x + 12y = 23 \\ x + 2y = 3 \end{cases}$

(g) $\begin{cases} 7u + 2v = 9 \\ 3u - 4v = 5 \end{cases}$ *(h) $\begin{cases} 12m + 5n = 13 \\ 5m + 6n = -13 \end{cases}$

3. Compute the value of each determinant.

*(a) $\begin{vmatrix} 5 & 2 & 1 \\ 1 & 4 & 0 \\ 3 & 9 & 7 \end{vmatrix}$ *(b) $\begin{vmatrix} 5 & 3 & -8 \\ -1 & 0 & 2 \\ 4 & 7 & 6 \end{vmatrix}$

*(c) $\begin{vmatrix} 1 & 2 & 3 \\ 4 & 5 & 6 \\ 7 & 8 & 9 \end{vmatrix}$ *(d) $\begin{vmatrix} 5 & 8 & 1 \\ 2 & 0 & 6 \\ 1 & 4 & 0 \end{vmatrix}$

(e) $\begin{vmatrix} 4 & 0 & 6 \\ 2 & -5 & 1 \\ 7 & 3 & 0 \end{vmatrix}$ *(f) $\begin{vmatrix} 1 & 2 & 5 \\ 3 & 0 & 4 \\ 0 & -7 & 6 \end{vmatrix}$

(g) $\begin{vmatrix} 9 & 8 & 5 \\ 4 & 3 & 7 \\ 1 & 6 & 2 \end{vmatrix}$ (h) $\begin{vmatrix} 7 & -4 & 5 \\ 8 & 3 & 0 \\ 2 & 6 & -9 \end{vmatrix}$

4. Solve each system of equations by using Cramer's rule.

*(a) $\begin{cases} x + 3y = 5 \\ x + z = 4 \\ 3x + 2y - z = 2 \end{cases}$ *(b) $\begin{cases} x + 2y + z = 0 \\ y = 6 \\ 2x + 4y + 5z = 9 \end{cases}$

*(c) $\begin{cases} x + y + z = 6 \\ 2x + 3y - z = 0 \\ 5x - 3y - 7z = -4 \end{cases}$ (d) $\begin{cases} x + 2y + 3z = 7 \\ 2x - 3y + 8z = -3 \\ -5x - 8y + 5z = 11 \end{cases}$

*(e) $\begin{cases} 5x + 2y - 3z = 10 \\ 2x + 7y + 5z = 9 \\ 3x - 3y + 2z = 8 \end{cases}$ *(f) $\begin{cases} 4r - 3s + 2t = 9 \\ 5r + s - 5t = -3 \\ 2r - 8s + t = 5 \end{cases}$

(g) $\begin{cases} 6x + 2y + z = 4 \\ 5x - 3y + 4z = 0 \\ x - 2y + 8z = 3 \end{cases}$ (h) $\begin{cases} 3x + 5y + 2z = 15 \\ 2x - 3y = 9 \\ 4x + 2y - 9z = 6 \end{cases}$

7.3 HIGHER-ORDER DETERMINANTS

We have already studied the evaluation and use of determinants of order 2, that is, determinants with two rows and two columns.

$$
\begin{array}{cc}
 & \text{col. 1}\quad\text{col. 2} \\
 & \downarrow\qquad\downarrow \\
\text{row 1} \rightarrow & \begin{vmatrix} a & b \\ c & d \end{vmatrix}
\end{array}
$$

Notation is sometimes used to specify the position of each number by its row and column, as

$$
\begin{vmatrix} a_{11} & a_{12} \\ a_{21} & a_{22} \end{vmatrix}
\qquad
\begin{cases}
a_{11} = \text{number in row 1, column 1} \\
a_{12} = \text{number in row 1, column 2} \\
a_{21} = \text{number in row 2, column 1} \\
a_{22} = \text{number in row 2, column 2}
\end{cases}
$$

We have also studied determinants of order 3, that is, determinants having three rows and three columns. Determinants of order 4 have four rows and four columns. Unfortunately, such determinants (and still-higher-order determinants) cannot be evaluated by the mechanical techniques used with determinants of orders 2 and 3. Solution of the general system of four linear equations in four unknowns by the addition or elimination method suggests there are 24 products in the definition of a determinant of order 4. (Recall that there are two products in a determinant of order 2 and six products in a determinant of order 3.) If our earlier mechanical techniques were applied to a determinant of order 4, only 8 of the 24 products would be produced: four from one direction and four from the opposite direction. The other 16 products are not readily available without a new definition of determinant.

First, let's redefine a determinant of order 3 and show that this definition leads to the same value as did the earlier definition.

$$
\begin{vmatrix} a_{11} & a_{12} & a_{13} \\ a_{21} & a_{22} & a_{23} \\ a_{31} & a_{32} & a_{33} \end{vmatrix}
= +a_{11}\begin{vmatrix} a_{22} & a_{23} \\ a_{32} & a_{33} \end{vmatrix}
- a_{12}\begin{vmatrix} a_{21} & a_{23} \\ a_{31} & a_{33} \end{vmatrix}
$$

$$
+ a_{13}\begin{vmatrix} a_{21} & a_{22} \\ a_{31} & a_{32} \end{vmatrix}
$$

The numbers a_{11}, a_{12}, and a_{13} are all from row 1. Each is multiplied by the determinant that remains when its row and column are deleted.

For a_{11}:
$$\begin{vmatrix} a_{11} & a_{12} & a_{13} \\ a_{21} & a_{22} & a_{23} \\ a_{31} & a_{32} & a_{33} \end{vmatrix} \longrightarrow +a_{11} \begin{vmatrix} a_{22} & a_{23} \\ a_{32} & a_{33} \end{vmatrix}$$

For a_{12}:
$$\begin{vmatrix} a_{11} & a_{12} & a_{13} \\ a_{21} & a_{22} & a_{23} \\ a_{31} & a_{32} & a_{33} \end{vmatrix} \longrightarrow -a_{12} \begin{vmatrix} a_{21} & a_{23} \\ a_{31} & a_{33} \end{vmatrix}$$

For a_{13}:
$$\begin{vmatrix} a_{11} & a_{12} & a_{13} \\ a_{21} & a_{22} & a_{23} \\ a_{31} & a_{32} & a_{33} \end{vmatrix} \longrightarrow +a_{13} \begin{vmatrix} a_{21} & a_{22} \\ a_{31} & a_{32} \end{vmatrix}$$

If the sum of the two subscripts $(i + j)$ is even, then the sign is $+$. This occurs for a_{11}, a_{13}, a_{22}, a_{31}, and a_{33}. If the sum $i + j$ is odd, then the sign is $-$. This occurs for a_{12}, a_{21}, a_{23}, and a_{32}. The signs alternate in going across any row or down any column. This can be illustrated symbolically as

$$\begin{matrix} + & - & + \\ - & + & - \\ + & - & + \end{matrix}$$
$\left\{\begin{matrix}\text{The pattern extends} \\ \text{indefinitely both across} \\ \text{and down.}\end{matrix}\right.$

Although the definition suggests using row 1 to obtain the smaller-order determinants, any row or column can be used. As an example, let's use column 2.

$$\begin{vmatrix} a_{11} & a_{12} & a_{13} \\ a_{21} & a_{22} & a_{23} \\ a_{31} & a_{32} & a_{33} \end{vmatrix} = -a_{12}\begin{vmatrix} a_{21} & a_{23} \\ a_{31} & a_{33} \end{vmatrix} + a_{22}\begin{vmatrix} a_{11} & a_{13} \\ a_{31} & a_{33} \end{vmatrix}$$

$$-\, a_{32}\begin{vmatrix} a_{11} & a_{13} \\ a_{21} & a_{23} \end{vmatrix}$$

Which row or column you select should depend on the numbers in the determinant. For example, if a row or column contains some zeros, selecting it will simplify computations.

The smaller determinants obtained in this process are called *minors*. Above, a_{12}'s minor is the determinant

$$\begin{vmatrix} a_{21} & a_{23} \\ a_{31} & a_{33} \end{vmatrix}$$

A signed minor ($+$ or $-$ according to its position) is called a *cofactor*. Thus, a_{12}'s cofactor is

$$-\begin{vmatrix} a_{21} & a_{23} \\ a_{31} & a_{33} \end{vmatrix}$$

Now let's use the new definition of determinant to evaluate the determinant

$$\begin{vmatrix} 1 & 2 & 4 \\ 3 & 7 & 5 \\ 9 & 6 & 8 \end{vmatrix}$$

The value of this determinant was shown to be -112, by earlier techniques. We will use row 1 because no other row or column seems to offer any advantage.

$$\begin{vmatrix} 1 & 2 & 4 \\ 3 & 7 & 5 \\ 9 & 6 & 8 \end{vmatrix} = 1\begin{vmatrix} 7 & 5 \\ 6 & 8 \end{vmatrix} - 2\begin{vmatrix} 3 & 5 \\ 9 & 8 \end{vmatrix} + 4\begin{vmatrix} 3 & 7 \\ 9 & 6 \end{vmatrix}$$

$$= 1(7 \cdot 8 - 6 \cdot 5) - 2(3 \cdot 8 - 9 \cdot 5) + 4(3 \cdot 6 - 9 \cdot 7)$$

$$= 1(56 - 30) - 2(24 - 45) + 4(18 - 63)$$

$$= 1(26) - 2(-21) + 4(-45)$$

$$= 26 + 42 - 180 = -112 \quad \checkmark$$

The new definition of determinants was given for determinants of order 3. The extension of this definition to higher-order determinants should be apparent. Here, for example, is the definition of a determinant of order 4.

$$\begin{vmatrix} a_{11} & a_{12} & a_{13} & a_{14} \\ a_{21} & a_{22} & a_{23} & a_{24} \\ a_{31} & a_{32} & a_{33} & a_{34} \\ a_{41} & a_{42} & a_{43} & a_{44} \end{vmatrix} = a_{11} \begin{vmatrix} a_{22} & a_{23} & a_{24} \\ a_{32} & a_{33} & a_{34} \\ a_{42} & a_{43} & a_{44} \end{vmatrix} - a_{12} \begin{vmatrix} a_{21} & a_{23} & a_{24} \\ a_{31} & a_{33} & a_{34} \\ a_{41} & a_{43} & a_{44} \end{vmatrix}$$

$$+ a_{13} \begin{vmatrix} a_{21} & a_{22} & a_{24} \\ a_{31} & a_{32} & a_{34} \\ a_{41} & a_{42} & a_{44} \end{vmatrix} - a_{14} \begin{vmatrix} a_{21} & a_{22} & a_{23} \\ a_{31} & a_{32} & a_{33} \\ a_{41} & a_{42} & a_{43} \end{vmatrix}$$

The definition is used below to evaluate a determinant of order 4. Column 1 has been selected, rather than row 1, however. Row 4 would also be a good choice.

$$\begin{vmatrix} 1 & 5 & 9 & 3 \\ 2 & 4 & 1 & 8 \\ 0 & 3 & 7 & 1 \\ -3 & 5 & 0 & 6 \end{vmatrix} = 1 \begin{vmatrix} 4 & 1 & 8 \\ 3 & 7 & 1 \\ 5 & 0 & 6 \end{vmatrix} - 2 \begin{vmatrix} 5 & 9 & 3 \\ 3 & 7 & 1 \\ 5 & 0 & 6 \end{vmatrix}$$

$$+ 0 - (-3) \begin{vmatrix} 5 & 9 & 3 \\ 4 & 1 & 8 \\ 3 & 7 & 1 \end{vmatrix}$$

$$= 1(-125) - 2(-12) + 3(-20)$$

$$= -161 \ \checkmark$$

EXERCISE SET C

1. Evaluate each determinant by use of cofactors.

*(a) $\begin{vmatrix} 1 & 0 & 2 \\ 4 & 1 & 6 \\ 9 & 7 & 8 \end{vmatrix}$

*(b) $\begin{vmatrix} 6 & 7 & 8 \\ 0 & -1 & 1 \\ 9 & 4 & 3 \end{vmatrix}$

*Answers to starred problems are to be found at the back of the book.

(c) $\begin{vmatrix} 0 & 1 & -2 \\ -4 & 0 & 3 \\ -5 & 6 & 0 \end{vmatrix}$

★(d) $\begin{vmatrix} 5 & 0 & 9 \\ 6 & 0 & -2 \\ -3 & 0 & 8 \end{vmatrix}$

(e) $\begin{vmatrix} 7 & 15 & -9 \\ 6 & -8 & 0 \\ 4 & 1 & 20 \end{vmatrix}$

(f) $\begin{vmatrix} 6 & -19 & -8 \\ -4 & 9 & 2 \\ 0 & -3 & 0 \end{vmatrix}$

★(g) $\begin{vmatrix} 5 & 2 & 1 & 0 \\ 9 & 8 & 7 & 6 \\ 4 & -3 & 7 & 4 \\ 8 & 1 & -1 & 6 \end{vmatrix}$

★(h) $\begin{vmatrix} 1 & 7 & 3 & -9 \\ 6 & 0 & 1 & 9 \\ 4 & 0 & -8 & 6 \\ 5 & -1 & 7 & 10 \end{vmatrix}$

(i) $\begin{vmatrix} 5 & 5 & 5 & 5 \\ -1 & 0 & 1 & 0 \\ 3 & -9 & 1 & 6 \\ 4 & 5 & 8 & -7 \end{vmatrix}$

(j) $\begin{vmatrix} 9 & 8 & 7 & 6 \\ 5 & 4 & 3 & 2 \\ 0 & 0 & 0 & 1 \\ -8 & -7 & -6 & -5 \end{vmatrix}$

2. Solve each system by using Cramer's rule.

★(a) $\begin{cases} w+3x+5y+2z= 11 \\ 2w-5x+3y-3z= 5 \\ 4w+7x+2y-5z=-3 \\ 3w+ x+ y+ z= 21 \end{cases}$

(b) $\begin{cases} w+ x+ y+ z= 3 \\ 2w +3y-7z= 20 \\ 5w-2x+ y-5z=-8 \\ -3w +2y+ z= 5 \end{cases}$

★(c) $\begin{cases} 8A+5B+6D+ F= 3 \\ 9A-7B-2D- F= 10 \\ -4A- B+5D+7F=-2 \\ A+4B+ D = 0 \end{cases}$

(d) $\begin{cases} 2a+9b+5c+8d= 1 \\ 10a+ b+2c-5d= 18 \\ -8a-5b-3c+ d=-6 \\ 3b+3c-8d= 12 \end{cases}$

★(e) $\begin{cases} r+2s+ t+ u=10 \\ 5r-9s+4t+8u= 0 \\ -3r+5s+ t-2u= 1 \\ 10r-7s-9t+5u= 4 \end{cases}$

(f) $\begin{cases} 8r + 5s - 3t - 4w = 4 \\ 5r + 6t + w = 0 \\ 7r + 3s + 5t + 8w = 14 \\ -9r - 8s + 2t - 5w = -16 \end{cases}$

7.4 MATRICES

A *matrix* is a rectangular array of numbers. The following are examples of matrices:

$$\begin{pmatrix} 2 & -1 \\ 3 & 4 \end{pmatrix} \quad \begin{pmatrix} 1 & 6 & 9 \\ 2 & 5 & 0 \end{pmatrix} \quad \begin{pmatrix} 5 \\ -4 \\ 17 \end{pmatrix}$$

Some matrices have the square shape of determinants, that is, the same number of rows as columns. On the other hand, matrices do not have to be square. The second matrix above has two rows and three columns; the third matrix has three rows and only one column. A matrix does not have a *single* value. It is an array of values. You can compute a value for a determinant, but you cannot compute the value of a matrix.

In this section, matrices will be used to solve systems of linear equations by a method called *Gauss' elimination* or *Gaussian elimination*. It is named for the German mathematician Karl Friedrich Gauss (1777–1855). The method is also referred to less formally as the "sweep-out" method.

The elimination method begins by placing the coefficients of the variables and the separate constants into a matrix. The variables are omitted, but the elements of the matrix are kept in order corresponding to their positions in the original equations. And it is assumed at the beginning that the system of equations is set up for the addition method, so that x terms align under x terms, etc. Here is an example of how a small system of linear equations is set up in a matrix for solution by Gaussian elimination.

$$\begin{cases} x + 2y = 1 \\ 2x + 7y = -4 \end{cases} \rightarrow \begin{pmatrix} 1 & 2 & 1 \\ 2 & 7 & -4 \end{pmatrix}$$

Next, a series of *elementary row operations* is performed on the matrix in order to produce a simpler matrix: specifically, one in which enough coefficients have been changed to zero that the values of the variables can be read from the matrix. For example, the preceding matrix can be reduced by elementary row operations to the matrix

$$\begin{pmatrix} 1 & 0 & 5 \\ 0 & 1 & -2 \end{pmatrix}$$

which is the same as the system

$$\begin{cases} 1x + 0y = 5 \\ 0x + 1y = -2 \end{cases}$$

or simply

$$\begin{cases} x = 5 \\ y = -2 \end{cases}$$

The elementary row operations used are:

1. Interchange two rows.
2. Replace any row by a multiple of itself.
3. Multiply any row by a number and add it (element by element) to another row in order to replace the latter row.

Keep in mind that the elementary row operations are being performed on the skeleton of an equation, so that in effect they are elementary equation operations. To interchange two equations in a system preserves the system. To multiply all members of an equation (both sides) by a constant does not change the solution of the system. Nor does multiplying an equation by a constant and adding it to another equation (thus replacing it) change the system.

If we let an asterisk (∗) represent any constant, then at first the matrix representation of the system appears as

$$\begin{pmatrix} * & * & * \\ * & * & * \end{pmatrix}$$

and we would like to change it to a matrix of the form

$$\begin{pmatrix} 1 & 0 & * \\ 0 & 1 & * \end{pmatrix}$$

To avoid problems, it is suggested that you follow the steps as outlined, so that your approach to all these problems will be consistent.

$$\begin{pmatrix} * & * & * \\ * & * & * \end{pmatrix}$$

$$\downarrow$$

$$\begin{pmatrix} 1 & * & * \\ * & * & * \end{pmatrix} \quad \begin{cases} \text{First, get a 1 in the top} \\ \text{left corner of the matrix.} \end{cases}$$

$$\downarrow$$

$$\begin{pmatrix} 1 & * & * \\ 0 & * & * \end{pmatrix} \quad \begin{cases} \text{Next, get a 0 in the position} \\ \text{below it} \end{cases}$$

$$\downarrow$$

$$\begin{pmatrix} 1 & * & * \\ 0 & 1 & * \end{pmatrix} \quad \begin{cases} \text{Third, get a 1 next to the 0} \\ \text{in row 2.} \end{cases}$$

$$\downarrow$$

$$\begin{pmatrix} 1 & 0 & * \\ 0 & 1 & * \end{pmatrix} \quad \begin{cases} \text{Finally, get a 0 above the} \\ \text{newly obtained 1.} \end{cases}$$

Example 6. *Use matrices and elementnry row operations to solve the system*

$$\begin{cases} x + 2y = 1 \\ 2x + 7y = -4 \end{cases}$$

The system can be written as the matrix

$$\begin{pmatrix} 1 & 2 & 1 \\ 2 & 7 & -4 \end{pmatrix}$$

First get a 1 in the top left position. And since there is a 1 there already, proceed to the next step. Get a 0 below the 1. This can be done by multiplying row 1 (each term) by -2 and adding it to row 2 to produce a new row 2. Multiplying row 1 by -2 produces

$$-2 \quad -4 \quad -2$$

When these numbers are added to

$$2 \quad 7 \quad -4$$

the new row two produced is

$$0 \quad 3 \quad -6$$

Thus, the matrix is now

$$\begin{pmatrix} 1 & 2 & 1 \\ 0 & 3 & -6 \end{pmatrix}$$

Next, get a 1 in the position next to the zero. Perform an operation on row 2 to change the 3 to a 1. Multiply the row by 1/3. The result is

$$\begin{pmatrix} 1 & 2 & 1 \\ 0 & 1 & -2 \end{pmatrix}$$

The last step is to get a 0 above the new 1. This can be accomplished by multiplying row 2 by −2 and adding it to row 1 in order to produce a new row 1. When row 2 is multiplied by −2, the result is

$$0 \quad -2 \quad 4$$

When this is added to

$$1 \quad 2 \quad 1$$

the new row one produced is

$$1 \quad 0 \quad 5$$

and the matrix becomes

$$\begin{pmatrix} 1 & 0 & 5 \\ 0 & 1 & -2 \end{pmatrix}$$

Now you can see from the matrix that x is 5 and y is −2.

If the matrix of Example 6 was instead

$$\begin{pmatrix} 4 & -3 & 7 \\ 1 & 5 & -8 \end{pmatrix}$$

then row operation 1, interchange of rows, will produce a matrix having 1 in the top left position. After interchanging rows 1 and 2, the matrix appears as

$$\begin{pmatrix} 1 & 5 & -8 \\ 4 & -3 & 7 \end{pmatrix}$$

If the original matrix was

$$\begin{pmatrix} 2 & 7 & 6 \\ 5 & 9 & 4 \end{pmatrix}$$

then you can multiply row 1 by $\frac{1}{2}$ to get a 1 in the top left position. The result is

$$\begin{pmatrix} 1 & \frac{7}{2} & 3 \\ 5 & 9 & 4 \end{pmatrix}$$

A system of three linear equations in three unknowns can be solved by matrices in a similar fashion. The system

$$\begin{cases} 7x + 9y - 4z = 19 \\ x + 3y + 2z = 7 \\ 3x + 7y + 11z = 8 \end{cases}$$

can be written in matrix form as

$$\begin{pmatrix} 7 & 9 & -4 & 19 \\ 1 & 3 & 2 & 7 \\ 3 & 7 & 11 & 8 \end{pmatrix}$$

Application of elementary row operations will eventually produce

$$\begin{pmatrix} 1 & 0 & 0 & -3 \\ 0 & 1 & 0 & 4 \\ 0 & 0 & 1 & -1 \end{pmatrix}$$

or $x = -3, y = 4, z = -1$. In general, the matrix

$$\begin{pmatrix} * & * & * & * \\ * & * & * & * \\ * & * & * & * \end{pmatrix}$$

is reduced to

$$\begin{pmatrix} 1 & 0 & 0 & * \\ 0 & 1 & 0 & * \\ 0 & 0 & 1 & * \end{pmatrix}$$

There are more steps because there is more to eliminate. But the pattern followed and the nature of the steps is essentially the same as before. Here, then, are the steps in sequence for the general case suggested above.

$$\begin{pmatrix} * & * & * & * \\ * & * & * & * \\ * & * & * & * \end{pmatrix}$$

$\downarrow$

$$\begin{pmatrix} 1 & * & * & * \\ * & * & * & * \\ * & * & * & * \end{pmatrix}$$
{First, get a 1 in the top left corner.

$\downarrow$

$$\begin{pmatrix} 1 & * & * & * \\ 0 & * & * & * \\ 0 & * & * & * \end{pmatrix}$$
{Then, get a 0 below it and another 0 below that zero. This is really two separate steps.

$\downarrow$

$$\begin{pmatrix} 1 & * & * & * \\ 0 & 1 & * & * \\ 0 & * & * & * \end{pmatrix}$$
{Next, get a 1 in row 2, column 2

$\downarrow$

$$\begin{pmatrix} 1 & 0 & * & * \\ 0 & 1 & * & * \\ 0 & 0 & * & * \end{pmatrix}$$
{Now, get a 0 above this new 1 and a 0 below it. This is two steps.

$\downarrow$

$$\begin{pmatrix} 1 & 0 & * & * \\ 0 & 1 & * & * \\ 0 & 0 & 1 & * \end{pmatrix}$$
{Next, get a 1 in row 3, column 3.

$\downarrow$

$$\begin{pmatrix} 1 & 0 & 0 & * \\ 0 & 1 & 0 & * \\ 0 & 0 & 1 & * \end{pmatrix}$$
{Finally, get 0's in both positions above this new 1. This is two more steps.

Example 7. *Use matrices and elementary row operations to solve the system*

$$\begin{cases} 7x + 9y - 4z = 19 \\ x + 3y + 2z = 7 \\ 3x + 7y + 11z = 8 \end{cases}$$

The system can be written as the matrix

$$\begin{pmatrix} 7 & 9 & -4 & 19 \\ 1 & 3 & 2 & 7 \\ 3 & 7 & 11 & 8 \end{pmatrix}$$

First, interchange rows 1 and 2 in order to place a 1 in the top left corner.

$$\begin{pmatrix} 1 & 3 & 2 & 7 \\ 7 & 9 & -4 & 19 \\ 3 & 7 & 11 & 8 \end{pmatrix}$$

Next, use this 1 to obtain 0's in the two positions below it. The zero in row 2, column 1, is obtained by multiplying row 1 by -7 and adding that to row 2. When row 1 is multiplied by -7, the result is

$$-7 \qquad -21 \qquad -14 \qquad -49$$

And when these elements are added to corresponding elements of row 2, that row becomes

$$0 \qquad -12 \qquad -18 \qquad -30$$

Similarly, multiply row 1 by -3 and add that to row 3 to produce a 0 in row 3, column 1. If the elements of row 1 are multiplied by -3, they become

$$-3 \qquad -9 \qquad -6 \qquad -21$$

And when they are added to the elements of row 3, the result is

$$0 \qquad -2 \qquad 5 \qquad -13$$

Shown next is the matrix with the new row 2 and new row 3 computed above.

$$
\begin{pmatrix}
1 & 3 & 2 & 7 \\
0 & -12 & -18 & -30 \\
0 & -2 & 5 & -13
\end{pmatrix}
$$

Next, obtain a 1 in row 2, column 2, by multiplying each element of that row by $-\frac{1}{12}$, the reciprocal of -12.

$$
\begin{pmatrix}
1 & 3 & 2 & 7 \\
0 & 1 & \frac{3}{2} & \frac{5}{2} \\
0 & -2 & 5 & -13
\end{pmatrix}
$$

Now get 0's above and below the new 1. A zero below it can be obtained by multiplying row 2 by 2 and adding it to row 3. When row 2 is multiplied by 2, the result is

$$0 \qquad 2 \qquad 3 \qquad 5$$

Add this to

$$0 \qquad -2 \qquad 5 \qquad -13$$

and the new row 3 obtained is

$$0 \qquad 0 \qquad 8 \qquad -8$$

Now multiply row 2 by -3 and add that result to row 1. That is, add

$$0 \qquad -3 \qquad -\frac{9}{2} \qquad -\frac{15}{2} \qquad (-3 \text{ times row 2})$$

to

$$1 \qquad 3 \qquad 2 \qquad 7$$

to get the new row 1:

$$1 \qquad 0 \qquad -\frac{5}{2} \qquad -\frac{1}{2}$$

The matrix is now

$$
\begin{pmatrix}
1 & 0 & -\frac{5}{2} & -\frac{1}{2} \\
0 & 1 & \frac{3}{2} & \frac{5}{2} \\
0 & 0 & 8 & -8
\end{pmatrix}
$$

Next, get a 1 in row 3, column 3. This is done by multiplying row 3 by $\frac{1}{8}$, the reciprocal of 8. The new matrix is

$$\begin{pmatrix} 1 & 0 & -\frac{5}{2} & -\frac{1}{2} \\ 0 & 1 & \frac{3}{2} & \frac{5}{2} \\ 0 & 0 & 1 & -1 \end{pmatrix}$$

Finally, get 0's above this new 1. A zero in row 1, column 3, can be obtained by multiplying row 3 by $\frac{5}{2}$ (the opposite of $-\frac{5}{2}$) and adding that result to row 1. That is, add

$$0 \qquad 0 \qquad \frac{5}{2} \qquad -\frac{5}{2}$$

to

$$1 \qquad 0 \qquad -\frac{5}{2} \qquad -\frac{1}{2}$$

to get the new row 1:

$$1 \qquad 0 \qquad 0 \qquad -3$$

A zero for row 2, column 3, is obtained by multiplying row 3 by $-\frac{3}{2}$ (the opposite of $\frac{3}{2}$) and adding that result to row 2. That is, add

$$0 \qquad 0 \qquad -\frac{3}{2} \qquad \frac{3}{2}$$

to

$$0 \qquad 1 \qquad \frac{3}{2} \qquad \frac{5}{2}$$

to get the new row 2:

$$0 \qquad 1 \qquad 0 \qquad 4$$

The final matrix is

$$\begin{pmatrix} 1 & 0 & 0 & -3 \\ 0 & 1 & 0 & 4 \\ 0 & 0 & 1 & -1 \end{pmatrix}$$

Thus, the system has been reduced to

$$\begin{cases} 1x + 0y + 0z = -3 \\ 0x + 1y + 0z = 4 \\ 0x + 0y + 1z = -1 \end{cases}$$

or simply

$$\begin{cases} x = -3 \\ y = 4 \quad \checkmark \\ z = -1 \end{cases}$$

EXERCISE SET D

1. Solve each system of linear equations by Gaussian elimination ("sweep out").

★(a) $\begin{cases} x + 3y = 10 \\ -3x + 2y = 3 \end{cases}$

(b) $\begin{cases} x - 2y = 3 \\ 4x + 5y = 51 \end{cases}$

(c) $\begin{cases} m - 5n = 23 \\ 2m - 3n = 11 \end{cases}$

★(d) $\begin{cases} 2x + 2y = 12 \\ x + 3y = 14 \end{cases}$

(e) $\begin{cases} 5x + 12y = 23 \\ x + 2y = 3 \end{cases}$

★(f) $\begin{cases} 2u + 3v = 5 \\ 5u - 4v = -22 \end{cases}$

★(g) $\begin{cases} 5x + 4y = 3 \\ 4x + 6y = 1 \end{cases}$

(h) $\begin{cases} 9x - 8y = 0 \\ 3x + 4y = 5 \end{cases}$

★(i) $\begin{cases} x + 2y + z = 6 \\ -x + y + 2z = 6 \\ z = 3 \end{cases}$

(j) $\begin{cases} x + 2y + z = 0 \\ y = 6 \\ 2x + 4y + 5z = 9 \end{cases}$

★(k) $\begin{cases} x - 3y + z = 4 \\ 3x - 8y + 3z = 13 \\ -2x + y + 2z = -5 \end{cases}$

(l) $\begin{cases} x - y - 4z = 12 \\ 2x + y + 3z = -6 \\ 5x + 2y + 2z = 1 \end{cases}$

(m) $\begin{cases} x + 2y + 3z = 7 \\ 2x - 3y + 8z = -3 \\ -5x - 8y + 5z = 11 \end{cases}$

★(n) $\begin{cases} 5x + 3y + 14z = 13 \\ x + y + 4z = 3 \\ -4x - 3y + 3z = 21 \end{cases}$

★(o) $\begin{cases} 3t - 4u + 3v = 9 \\ 4t - 6u + 10v = 28 \\ t - 2u + 5v = 15 \end{cases}$

(p) $\begin{cases} 2p - 4q + 10r = -6 \\ 6p - 10q + 12r = -32 \\ -5p - 9q + 3r = 5 \end{cases}$

7.5 APPLICATION: CIRCLES

The equation of the circle which passes through three given points can be obtained by establishing and solving a system of three linear equations in three unknowns. Any circle can be written in the form

$$x^2 + y^2 + ax + by + c = 0$$

If the x and y coordinates of a point are substituted into this equation, there will be three unknowns: a, b, c. If this is done for three different points, then three equations in three unknowns will be generated.

Suppose that a circle passes through the points $(-2, 6)$, $(3, 1)$, and $(1, -3)$. When we use the point $(-2, 6)$ and substitute -2 for x and 6 for y into

*Answers to starred problems are to be found at the back of the book.

$$x^2 + y^2 + ax + by + c = 0$$

we get

$$4 + 36 + a(-2) + b(6) + c = 0$$

or

$$-2a + 6b + c = -40$$

Similarly, the point (3, 1) yields

$$3a + b + c = -10$$

and (1, −3) yields

$$a - 3b + c = -10$$

The system is

$$\begin{cases} -2a + 6b + c = -40 \\ 3a + b + c = -10 \\ a - 3b + c = -10 \end{cases}$$

The solution to this system, by any of the techniques studied earlier in this chapter, is

$$a = \quad 4$$
$$b = \quad -2$$
$$c = -20$$

so

$$x^2 + y^2 + ax + by + c = 0$$

becomes

$$x^2 + y^2 + 4x - 2y - 20 = 0 \quad \checkmark$$

This is the equation of the circle through the points (−2, 6), (3, 1), and (1, −3).

7.6 APPLICATION: PARTIAL FRACTIONS

It is sometimes necessary in calculus to write a fraction as the sum of two or more fractions called *partial fractions*. A fraction such as

$$\frac{6x^2 - 55x - 21}{x(x + 1)(x - 7)}$$

can be written as the sum of three fractions. Each partial fraction has as its denominator one of the factors from the denominator of the original fraction.

$$\frac{6x^2 - 55x - 21}{x(x + 1)(x - 7)} = \frac{A}{x} + \frac{B}{x + 1} + \frac{C}{x - 7}$$

Here A, B, and C are constants. The values of A, B, and C can be determined as follows. First, combine the fractions that are on the right side. Then, since the denominators of the two equal fractions (the original and the new one produced on the right side) are the same, their numerators must be equal. So equate the numerators. This leads to three equations in A, B, and C, which can be solved by any of the methods studied earlier in the chapter. Here are the steps. First, combine the fractions that are on the right side.

$$\frac{6x^2 - 55x - 21}{x(x + 1)(x - 7)} = \frac{A(x + 1)(x - 7) + B(x)(x - 7) + C(x)(x + 1)}{x(x + 1)(x - 7)}$$

$$= \frac{Ax^2 - 6Ax - 7A + Bx^2 - 7Bx + Cx^2 + Cx}{x(x + 1)(x - 7)}$$

$$= \frac{(A + B + C)x^2 + (-6A - 7B + C)x - 7A}{x(x + 1)(x - 7)}$$

Since the denominators of the fractions are equal, the numerators of these two equal fractions must be equal. That is,

$$6x^2 - 55x - 21 = (A + B + C)x^2 + (-6A - 7B + C)x - 7A$$

This means that

$$6x^2 = (A + B + C)x^2$$
$$-55x = (-6A - 7B + C)x$$
$$-21 = -7A$$

Equating the coefficients of x^2 and x and the constants yields a system of three equations in three unknowns.

$$\begin{cases} A + B + C = 6 \\ -6A - 7B + C = -55 \\ -7A = -21 \end{cases}$$

The solution is

$$A = 3$$
$$B = 5$$
$$C = -2$$

Thus,

$$\frac{6x^2 - 55x - 21}{x(x + 1)(x - 7)} = \frac{3}{x} + \frac{5}{x + 1} - \frac{2}{x - 7}$$

Obtaining partial fractions for more complicated denominators is a little more involved. The partial fraction corresponding to a quadratic factor in the denominator has a linear numerator of the form $Ax + B$. If the nth power of any factor appears in the denominator of the original fraction, it must be written n times (using n separate fractions) in the partial-fraction representation. Furthermore, in that representation it is written to the nth power in the first denominator, to the $(n - 1)$st power in the second denominator, and so on until it is written as the first power in the last fraction. Here is an example:

$$\frac{7x^5 + 5x^4 - x^2 + 12}{(x + 3)^3(x^2 - 3x + 7)^2(x - 2)(x^2 + 5x + 1)}$$

$$= \frac{A}{(x + 3)^3} + \frac{B}{(x + 3)^2} + \frac{C}{x + 3} + \frac{Dx + E}{(x^2 - 3x + 7)^2}$$

$$+ \frac{Fx + G}{x^2 - 3x + 7} + \frac{H}{x - 2} + \frac{Ix + J}{x^2 + 5x + 1}$$

EXERCISE SET E

1. In each problem, determine the equation of the circle which passes through the three given points.
   ⋆(a) $(3, 3)$, $(8, -2)$, $(7, 1)$
   ⋆(b) $(2, 2)$, $(9, -5)$, $(2, -12)$
   (c) $(2, 1)$, $(-2, 3)$, $(3, -2)$

2. Write each fraction as the sum of two or more partial fractions. Factor denominator expressions whenever possible.
   ⋆(a) $\dfrac{9x + 10}{x(x + 5)}$ (b) $\dfrac{-2x + 34}{(x + 3)(x - 2)}$

   (c) $\dfrac{3x^2 + 3x - 16}{x^3 + 3x^2 - 4x}$ ⋆(d) $\dfrac{7x^2 - 82x + 279}{(x + 5)(x - 3)(x - 7)}$

   ⋆(e) $\dfrac{11x^2 + 33x + 2}{(x + 2)(x^2 + 5x + 1)}$ ⋆(f) $\dfrac{6x^2 - 3x - 54}{(x + 3)(x^2 - 4x - 12)}$

   (g) $\dfrac{-x^2 + 22x - 9}{x^3 + 3x^2 - 9x + 5}$ (h) $\dfrac{5x^3 + x^2 + 5x + 5}{(x^2 + 1)^2}$

⋆Answers to starred problems are to be found at the back of the book.

7.7 NONLINEAR SYSTEMS

The methods of substitution and elimination can be applied to systems involving quadratic equations and others. Whenever convenient, sketches should be made of the relations in the system. Solutions obtained must be checked in all equations of the system, since extraneous roots may appear.

Example 8. *Solve the system*

$$\begin{cases} x^2 + y^2 = 16 \\ x + y = 3 \end{cases}$$

The second equation can be written as $y = 3 - x$, after which $3 - x$ can be substituted for y in the first equation. When this is done, the first equation becomes

$$x^2 + (3 - x)^2 = 16$$
$$x^2 + 9 - 6x + x^2 = 16$$

or

$$2x^2 - 6x - 7 = 0$$

By using the quadratic formula, we get

$$x = \frac{6 \pm \sqrt{92}}{4} = \frac{2(3 \pm \sqrt{23})}{2 \cdot 2} = \frac{3 \pm \sqrt{23}}{2}$$

In order to now find y, substitute these values of x into the linear equation, because it will yield one y for each x and hence the two points of intersection (anticipated from the graph).

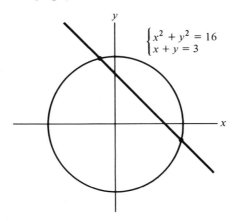

On the other hand, if these values of x are substituted into the quadratic equation, the result will be two y values for each x value. In other words, using the first equation will suggest that the line crosses the circle in four points. However, when those points are checked in the second equation, only two will check correctly. Using

$$x = \frac{3 + \sqrt{23}}{2}$$

in $y = 3 - x$ yields

$$y = 3 - \frac{3 + \sqrt{23}}{2} = \frac{6}{2} - \frac{3 + \sqrt{23}}{2} = \frac{3 - \sqrt{23}}{2}$$

So one point of intersection is

$$\left(\frac{3 + \sqrt{23}}{2}, \frac{3 - \sqrt{23}}{2} \right) \quad \checkmark$$

Using

$$x = \frac{3 - \sqrt{23}}{2}$$

in $y = 3 - x$ yields

$$y = 3 - \frac{3 - \sqrt{23}}{2} = \frac{6}{2} - \frac{3 - \sqrt{23}}{2} = \frac{3 + \sqrt{23}}{2}$$

The other point of intersection is

$$\left(\frac{3 - \sqrt{23}}{2}, \frac{3 + \sqrt{23}}{2} \right) \quad \checkmark$$

Example 9. *Solve the system*

$$\begin{cases} x^2 + y^2 = 2 \\ y^2 - x^2 = 4 \end{cases}$$

If the second equation is rewritten as $-x^2 + y^2 = 4$ and added to the first equation, the result is $2y^2 = 6$ or $y = \pm\sqrt{3}$. When $\sqrt{3}$ or $-\sqrt{3}$ is substituted for y into either $x^2 + y^2 = 2$ or $y^2 - x^2 = 4$, the result is $x = \pm i$. This means that there is no real solution. Graphically, there is no point in the xy plane (the real plane) where the circle and the hyperbola above meet. The graph verifies this.

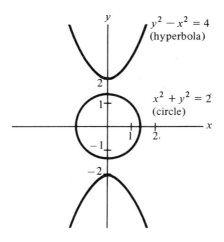

Example 10. *Solve the system*

$$\begin{cases} y = x^2 + 1 \\ x^2 + 4y^2 = 4 \end{cases}$$

Substitute $x^2 + 1$ for y into the second equation. The result is

$$x^2 + 4(x^2 + 1)^2 = 4$$
$$x^2 + 4x^4 + 8x^2 + 4 = 4$$
$$4x^4 + 9x^2 = 0$$
$$x^2(4x^2 + 9) = 0$$

Finally,

$$x = 0 \qquad x = \pm\sqrt{-\frac{9}{4}} = \pm\frac{3}{2}i$$

from which we conclude $x = 0$, since x must be real. Then

$$y = x^2 + 1 = 0 + 1 = 1$$

The curves meet at (0, 1) only, as shown in the graph at the top of the next page.

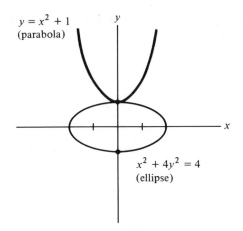

$y = x^2 + 1$
(parabola)

$x^2 + 4y^2 = 4$
(ellipse)

Example 11. *Write $\sqrt{2i}$ in $a + bi$ form.*

This apparently unrelated problem can indeed be solved by using the techniques of this section. To write $\sqrt{2i}$ in $a + bi$ form, square both sides of

$$\sqrt{2i} = a + bi$$

to obtain

$$2i = a^2 - b^2 + 2abi$$

If the real and imaginary parts are equated, the result is the system

$$\begin{cases} 2ab = 2 \\ a^2 - b^2 = 0 \end{cases}$$

Next, write $2ab = 2$ as $b = 2/(2a) = 1/a$ and substitute $1/a$ for b in the equation $a^2 - b^2 = 0$. The result is

$$a^2 - \left(\frac{1}{a}\right)^2 = 0$$

or

$$a^4 - 1 = 0$$

This equation can be factored and solved.

$$(a^2 + 1) \quad (a^2 - 1) = 0$$

$a^2 + 1 = 0$	$a^2 - 1 = 0$
$a = \pm i$	$a = \pm 1$

Since a must be real, $a = +1$ or $a = -1$. If $a = +1$, then $b = 1/a = 1$ also, and

$$\sqrt{2i} = 1 + i \quad \checkmark$$

If $a = -1$, then $b = -1$ also, and

$$\sqrt{2i} = -1 - i \quad \checkmark$$

There are two values for $\sqrt{2i}$: $1 + i$ and $-1 - i$.

EXERCISE SET F

1. Solve each system, if possible.

*(a) $\begin{cases} x^2 + y^2 = 25 \\ y - x = 1 \end{cases}$

*(b) $\begin{cases} x^2 + 4y^2 = 4 \\ x - y = -4 \end{cases}$

(c) $\begin{cases} 2x - y = 1 \\ x^2 + y^2 = 12 \end{cases}$

*(d) $\begin{cases} y = x^2 + 2 \\ 4x^2 + 9y^2 = 36 \end{cases}$

(e) $\begin{cases} y = -x^2 + 2 \\ 4x^2 + 9y^2 = 36 \end{cases}$

*(f) $\begin{cases} x^2 + y^2 = 16 \\ x^2 - y^2 = 16 \end{cases}$

(g) $\begin{cases} x^2 + y^2 = 16 \\ x^2 - y^2 = 1 \end{cases}$

*(h) $\begin{cases} x^2 + y^2 = 9 \\ 4x^2 + 16y^2 = 64 \end{cases}$

*(i) $\begin{cases} y = x^2 - 2x - 3 \\ y = -x^2 - x + 6 \end{cases}$

(j) $\begin{cases} x - y^2 = 0 \\ y^2 - x^2 = 16 \end{cases}$

*2. Write $\sqrt{3 + 4i}$ in $a + bi$ form.

*3. Where does the line with slope 1 and y intercept -3 intersect the circle $x^2 + y^2 - 6x + 2y + 3 = 0$?

4. A common problem of integral calculus is to find the area between two curves, that is, the area bounded by the curves. In order to set up the calculus, it is necessary to determine the points of intersection of the two curves involved. Also useful is a sketch that shows the desired area shaded. For each pair of equations, use algebra to determine the points of intersection. Then sketch the curves and shade the area between the curves.

*(a) $y = x^2 - 1$ and $y = x + 1$
   (b) $y = x - 7$ and $x = y^2 + 1$
*(c) $x^2 = 4 + y$ and $x^2 + y = 6$
   (d) $y = x^4$ and $y = x^2$
*(e) $y = x^3 - 6x^2 + 5x$ and $y = 0$

*Answers to starred problems are to be found at the back of the book.

chapter eight

---

sequences, induction, and iteration

8.1 SEQUENCE FUNCTIONS

Here are the first five numbers of a sequence.

$$2, 4, 6, 8, 10, \ldots$$

Assuming the obvious, the next numbers are 12, 14, 16, etc. This is a sequence of even numbers. The sequence above is a function. The numbers 2, 4, 6, 8, 10, etc., of the sequence constitute the range of the function. The domain is the set of positive integers. They specify position. The rule of correspondence can be illustrated as

Domain		Range	
1	$\xrightarrow{f}$	2	(The 1st number is 2.)
2	$\xrightarrow{f}$	4	(The 2nd number is 4.)
3	$\xrightarrow{f}$	6	(The 3rd number is 6.)
4	$\xrightarrow{f}$	8	(The 4th number is 8.)
5	$\xrightarrow{f}$	10	(The 5th number is 10.)

The rule of correspondence can be stated as

$$f(n) = 2n$$

The sequence $5, 7, 9, 11, 13, \ldots$ is described by the formula $f(n) = 2n + 3$. The sequence $0, 1, 2, 3, 4, \ldots$ is the function $f(n) = n - 1$. The sequence $\frac{1}{4}, \frac{1}{9}, \frac{1}{16}, \frac{1}{25}$, is the function $f(n) = 1/(n + 1)^2$.

Traditionally, a_n is used to mean $f(n)$. Thus, a_1 is the same as $f(1)$, the first term of the sequence; a_2 is the second term, and a_n is the nth term of a sequence. Similarly, rules of correspondence are often written without function notation. Thus, $f(n) = 2n$ can be written as $a_n = 2n$.

An *arithmetic progression* (AP) is a sequence in which the difference, called d, between any two adjacent terms is constant. The sequence $5, 7, 9, 11, \ldots$ is an arithmetic progression with $d = 2$. The fifth (next) term in the sequence is 2 more (d more) than the fourth term. $a_5 = a_4 + d = 11 + 2 = 13$. Also, $a_5 = a_1 + 4d$, since each difference moves you one position to the right in the sequence. In general,

$$\boxed{a_n = a_1 + (n - 1)d}$$

If in an AP $a_1 = 5$ and $d = 3$, then a_{19} can be found as

$$\begin{aligned} a_{19} &= a_1 + 18d \\ &= 5 + 18(3) \\ &= 59 \quad \checkmark \end{aligned}$$

The sum S_n of the first n terms of an arithmetic progression will now be shown to be

$$\boxed{S_n = \frac{n}{2}(a_1 + a_n)}$$

By S_n is meant $a_1 + a_2 + a_3 + \cdots + a_n$. That is,

$$S_n = a_1 + a_2 + a_3 + \cdots + a_n$$

If all terms on the right side are written using a_1 and d only, then S_n becomes

$$S_n = a_1 + (a_1 + d) + (a_1 + 2d) + \cdots + (a_1 + (n - 1)d) \quad (*)$$

Now reverse the order of the original S_n form, to get

$$S_n = a_n + a_{n-1} + a_{n-2} + \cdots + a_1$$

Next, write each term on the right side of this expression using a_n and d only. This result is

$$S_n = a_n + (a_n - d) + (a_n - 2d) + \cdots + (a_n - (n-1)d) \quad (**)$$

Now add the two forms of S_n labeled (*) and (**).

$$S_n = a_1 + a_1 + d + a_1 + 2d + \cdots + a_1 + (n-1)d \quad (*)$$
$$+ \quad S_n = a_n + a_n - d + a_n - 2d + \cdots + a_n - (n-1)d \quad (**)$$
$$\overline{2S_n = na_1 + na_n}$$

Note that all the d terms above add to zero and that all that remains on the right side is n a_1's and n a_n's. If n is factored from each term on the right side, the result is

$$2S_n = n(a_1 + a_n)$$

Finally, divide both sides by 2 to produce the desired result.

$$S_n = \frac{n}{2}(a_1 + a_n) \quad \checkmark$$

As an example, let's find the sum of the first 20 terms of the sequence $3, 7, 11, 15, \ldots$. It is an AP with $d = 4$ and $a_1 = 3$. Thus,

$$S_{20} = \frac{20}{2}(a_1 + a_{20})$$

or

$$S_{20} = 10(3 + a_{20})$$

Once a_{20} is determined, the computation can be completed.

$$a_{20} = a_1 + 19d$$
$$= 3 + 19(4)$$
$$= 79$$

Now

$$S_{20} = 10(3 + 79) = 820 \quad \checkmark$$

Finally, a formula for a_n for a specific AP can be determined by using the general formula $a_n = a_1 + (n - 1)d$. You merely supply a_1 and d. For example, consider an AP in which the first term is 7 and the difference is 2. Then

$$a_n = a_1 + (n - 1)d$$

becomes

$$a_n = 7 + (n - 1)2$$

or

$$a_n = 2n + 5 \quad \checkmark$$

In an arithmetic progression, a constant is added to a term to get the next term in the sequence. In a *geometric progression*, a constant is multiplied by a term to get the next term in the sequence. The constant is called r, for *ratio*. In the sequence 2, 6, 18, 54, . . . , each term is multiplied by 3 to get the next term to the right of it. So $r = 3$ for this geometric progression. The next term in the sequence is $a_5 = a_4 \cdot r = 54 \cdot 3 = 162$. The term after that is $a_6 = 486$. The number r is called the *ratio* because it is the quotient (or ratio) of any term divided by the term preceding it in the sequence.

Suppose that a geometric progression (GP) has $a_1 = 3$ and $r = 2$. Then, $a_2 = a_1 \cdot r = 3 \cdot 2 = 6$; $a_3 = a_2 \cdot r = 6 \cdot 2 = 12$; and so on. The term a_3 can also be computed as $a_3 = a_1 \cdot r^2$, since $a_3 = a_2 \cdot r = a_1 r \cdot r = a_1 r^2$. Similarly, $a_4 = a_1 r^3$, and in general

$$\boxed{a_n = a_1 r^{n-1}}$$

Example 1. *Find a_8 for a GP in which a_1 is 9 and $r = -2$.*

$$a_8 = a_1 r^7$$
$$= 9(-2)^7$$
$$= 9(-128)$$
$$= -1152 \quad \checkmark$$

The formula for computing a_n from a_1 and r can be used to obtain the formula for the terms of a GP. Consider the sequence 7, 21, 63, 189, ... Here $a_1 = 7$ and $r = \frac{21}{7} = 3$. When these two numbers are substituted into the form

$$a_n = a_1 \cdot r^{n-1}$$

the result is

$$a_n = 7 \cdot 3^{n-1}$$

which is a formula that can be used to obtain any term of the sequence.

The sum, S_n, of the first n terms of a geometric progression will now be shown to be

$$S_n = \frac{a_1(1 - r^n)}{1 - r} \qquad r \neq 1$$

The sum of the first n terms of a geometric sequence is

$$S_n = a_1 + a_2 + a_3 + \cdots + a_n$$

or

$$S_n = a_1 + a_1 r + a_1 r^2 + \cdots + a_1 r^{n-1} \qquad (*)$$

if each term is written in terms of a_1 and r only. Now multiply both sides of this equation by r. The result is

$$r S_n = a_1 r + a_1 r^2 + a_1 r^3 + \cdots + a_1 r^n \qquad (**)$$

Now align corresponding terms of (*) and (**) and subtract (**) from (*); that is,

$$S_n = a_1 + a_1 r + a_1 r^2 + \cdots + a_1 r^{n-1}$$
$$-r S_n = \quad - a_1 r - a_1 r^2 - \cdots - a_1 r^{n-1} - a_1 r^n$$
$$\overline{S_n - r S_n = a_1 - a_1 r^n}$$

or

$$(1 - r)S_n = a_1(1 - r^n)$$

after factoring out S_n on the left and a_1 on the right. Finally, divide both sides by $(1 - r)$ to get

$$S_n = \frac{a_1(1 - r^n)}{1 - r} \quad \checkmark$$

Example 2. *Find the sum of the first eight terms of the sequence*

$$5, 10, 20, \ldots$$

This is a geometric progression with a_1 given as 5 and $r = \frac{10}{5} = 2$. Thus,

$$S_8 = \frac{a_1(1 - r^8)}{1 - r} = \frac{5(1 - 2^8)}{1 - 2} = \frac{5(-255)}{-1} = 1275 \quad \checkmark$$

Consider what happens to the value of S_n when r is a fraction of magnitude less than 1 and n becomes very large. In

$$S_n = \frac{a_1(1 - r^n)}{1 - r}$$

the term r^n becomes negligibly small if $|r| < 1$ and n is very large. For example,

$$(\tfrac{1}{2})^{20} = \frac{1}{1,048,576}$$

More specifically, as n gets infinitely large, the value of r^n gets very close to zero and can be ignored. Thus, as

$$n \longrightarrow \infty$$
$$r^n \longrightarrow 0 \qquad (\text{for } |r| < 1)$$

The sum S_n under such conditions is denoted S_∞ and can be thought of as the sum of "all" the terms of the geometric sequence—the "sum to infinity." Thus,

$$S_\infty = S_n \Big|_{n \to \infty} = \frac{a_1(1 - r^n)}{1 - r} = \frac{a_1(1 - 0)}{1 - r} = \frac{a_1}{1 - r}$$

The result is

$$\boxed{S_\infty = \frac{a_1}{1 - r}} \qquad \text{for } |r| < 1$$

You can think of S_∞ as the number which the value of S_n gets closer and closer to as more and more terms of the sequence are added.

Consider the sequence

$$1, \frac{1}{2}, \frac{1}{4}, \frac{1}{8}, \frac{1}{16}, \frac{1}{32}, \ldots$$

$$S_1 = 1$$
$$S_2 = 1\tfrac{1}{2}$$
$$S_3 = 1\tfrac{3}{4}$$
$$S_4 = 1\tfrac{7}{8}$$
$$S_5 = 1\tfrac{15}{16}$$
$$S_6 = 1\tfrac{31}{32}$$

The sums are getting larger and appear to be closing in on 2. You would probably guess that $S_\infty = 2$. Here is the verification.

$$S_\infty = \frac{a_1}{1-r} = \frac{1}{1-\frac{1}{2}} = \frac{1}{\frac{1}{2}} = 2 \quad \checkmark$$

If $a_1, a_2, a_3, a_4, \ldots$ is a sequence, then the expression $a_1 + a_2 + a_3 + a_4 + \cdots$ is called an *infinite series*. Calculus techniques can be used to show that exponential, logarithmic, and trigonometric functions can be expressed as special series called *power series*. Values for tables of logarithms, sines, and cosines are calculated by using the first several terms of such series. Here are some examples of series representations of functions. Note that 3! is read "3-factorial" and means $1 \cdot 2 \cdot 3$, or 6. Similarly, $5! = 1 \cdot 2 \cdot 3 \cdot 4 \cdot 5$, or 120.

$$\sin x = x - \frac{x^3}{3!} + \frac{x^5}{5!} - \frac{x^7}{7!} + \frac{x^9}{9!} - \cdots$$

$$\cos x = 1 - \frac{x^2}{2!} + \frac{x^4}{4!} - \frac{x^6}{6!} + \frac{x^8}{8!} - \cdots \qquad \{x \text{ is in } \textit{radians}$$

$$e^x = 1 + x + \frac{x^2}{2!} + \frac{x^3}{3!} + \frac{x^4}{4!} + \cdots$$

$$\ln x = \left(\frac{x-1}{x}\right) + \frac{1}{2}\left(\frac{x-1}{x}\right)^2 + \frac{1}{3}\left(\frac{x-1}{x}\right)^3 + \cdots$$

An interesting relationship between the irrational number e and the trigonometric functions sine and cosine can be obtained when ix is used as the exponent in the series for e^x,

$$e^x = 1 + x + \frac{x^2}{2!} + \frac{x^3}{3!} + \frac{x^4}{4!} + \frac{x^5}{5!} + \cdots$$

Consider

$$e^{ix} = 1 + ix + \frac{(ix)^2}{2!} + \frac{(ix)^3}{3!} + \frac{(ix)^4}{4!} + \frac{(ix)^5}{5!} + \cdots$$

$$= 1 + ix + \frac{i^2 x^2}{2!} + \frac{i^3 x^3}{3!} + \frac{i^4 x^4}{4!} + \frac{i^5 x^5}{5!} + \cdots$$

$$= 1 + ix - 1 \cdot \frac{x^2}{2!} - i \cdot \frac{x^3}{3!} + 1 \cdot \frac{x^4}{4!} + i \cdot \frac{x^5}{5!} - \cdots$$

This series can be separated into real and imaginary parts.

$$e^{ix} = \left(1 - \frac{x^2}{2!} + \frac{x^4}{4!} - \cdots\right) + i\left(x - \frac{x^3}{3!} + \frac{x^5}{5!} - \cdots\right)$$

The real part is the series for cos x. The imaginary part is the series for sin x. Thus,

$$e^{ix} = \cos x + i \sin x$$

This relationship is called *Euler's formula*. It was introduced by Leonhard Euler, a famous Swiss mathematician of the eighteenth century. Since it involves sine and cosine, θ is often used instead of x.

$$\boxed{e^{i\theta} = \cos \theta + i \sin \theta}$$

Example 3. *Approximate the value of e by summing the first five terms of e^x with $x = 1$.*

Since

$$e^x = 1 + x + \frac{x^2}{2!} + \frac{x^3}{3!} + \frac{x^4}{4!} + \cdots$$

then

$$e^1 = 1 + 1 + \frac{1}{2!} + \frac{1}{3!} + \frac{1}{4!} + \cdots$$

or

$$e \doteq 1 + 1 + \frac{1}{2} + \frac{1}{6} + \frac{1}{24}$$
$$\doteq 1 + 1 + .5 + .16666 + .04166$$
$$\doteq 2.70832$$

The value of e correct to five decimal places is 2.71828. Our estimate can be improved by adding additional terms of the series to our sum. As it stands now, it is off by only .01.

EXERCISE SET A

1. Find a formula for a_n for each sequence.

⋆(a) 2, 3, 4, 5, 6, . . .   ⋆(b) 3, 6, 9, 12, . . .

⋆(c) $1, \frac{1}{2}, \frac{1}{3}, \frac{1}{4}, \ldots$ ⋆(d) $-2, -4, -6, -8, \ldots$

⋆Answers to starred problems are to be found at the back of the book.

(e) $\dfrac{1}{2}, \dfrac{2}{3}, \dfrac{3}{4}, \ldots$ (f) $1, \dfrac{1}{4}, \dfrac{1}{9}, \dfrac{1}{16}, \ldots$

$\star$(g) $9, 16, 25, 36, \ldots$ (h) $\sqrt{3}, \sqrt{4}, \sqrt{5}, \ldots$

$\star$(i) $\dfrac{5}{1}, \dfrac{5}{4}, \dfrac{5}{9}, \dfrac{5}{16}, \ldots$ (j) $0, 1, 8, 27, 64, \ldots$

2. Generate the first four terms of each sequence.

$\star$(a) $a_n = 3n + 1$ $\star$(b) $a_n = 2n - 5$

(c) $a_n = 3(n - 2)$ $\star$(d) $a_n = 3n^2 - 4$

(e) $a_n = \dfrac{n^2 + 2}{3}$ (f) $a_n = \dfrac{2n + 1}{3}$

$\star$(g) $a_n = \dfrac{2n(n - 1)}{3}$ $\star$(h) $a_n = \dfrac{n^2}{n + 1}$

(i) $a_n = \dfrac{n^2}{2} - n$ (j) $a_n = \dfrac{n}{3} - \dfrac{n^2}{5}$

3. For each arithmetic progression, find whatever is requested.

$\star$(a) $3, 5, 7, 9, \ldots$ Find a_{23}.

$\star$(b) $3, 5, 7, 9, \ldots$ Find S_{23}.

$\star$(c) $7, 10, 13, 16, \ldots$ Find a_{71}.

(d) $17, 15, 13, 11, 9, \ldots$ Find a_{41}.

$\star$(e) $2, 6, 10, 14, 18, \ldots$ Find S_{60}.

$\star$(f) $a_{12} = 64, d = 8$ Find a_1.

(g) $a_1 = 50, a_2 = 100$ Find a_8.

$\star$(h) $5, 8, 11, 14, \ldots$ Find a_{32} and S_{32}.

(i) $a_5 = 130, a_9 = 102$ Find a_1.

(j) $a_5 = 10, a_8 = 46$ Find a_1 and a_2.

4. For each geometric progression, find whatever is requested.

$\star$(a) $a_1 = 5, r = 2$ Find a_8.

$\star$(b) $4, 12, 36, \ldots$ Find a_7.

$\star$(c) $6, 3, \dfrac{3}{2}, \ldots$ Find a_{10}.

$\star$(d) $a_1 = 1, a_2 = 10$ Find a_9.

$\star$(e) $1, \dfrac{1}{3}, \dfrac{1}{9}, \dfrac{1}{27}, \ldots$ Find S_7.

(f) $\dfrac{3}{7}, \dfrac{3}{14}, \dfrac{3}{28}, \ldots$ Find the sum of the first 10 terms.

(g) $1, -3, 9, \ldots$ Find a_7.

*(h) 384, $-192, \ldots$ Find the sum of the first 5 terms.

(i) 320, 160, 80, $\ldots$ Find S_8.

*(j) $a_6 = 5$, $a_{11} = 1215$ Find r.

*(k) $a_2 = 40$, $a_5 = 2560$ Find a_3.

(l) $a_5 = 40$, $a_9 = 640$ Find a_1.

*(m) $a_2 = 12$, $a_6 = 972$ Find two different formulas for a_n.

(n) $a_4 = 8$, $a_7 = 1$ Find a formula for a_n.

5. Find S_∞ for each geometric sequence for which it exists.

*(a) 4, 2, 1, $\ldots$ *(b) $1, \dfrac{1}{3}, \dfrac{1}{9}, \dfrac{1}{27}, \ldots$

(c) $2, 1, \dfrac{1}{2}, \dfrac{1}{4}, \ldots$ (d) $3, \dfrac{3}{2}, \dfrac{3}{4}, \dfrac{3}{8}, \ldots$

(e) $\dfrac{1}{2}, \dfrac{1}{6}, \dfrac{1}{18}, \dfrac{1}{54}, \ldots$ *(f) $\dfrac{1}{8}, \dfrac{2}{24}, \dfrac{4}{72}, \ldots$

(g) 9, 6, 4, $\ldots$ *(h) 49, 7, $\ldots$

*(i) $\dfrac{3}{8}, \dfrac{3}{4}, \dfrac{3}{2}, \ldots$ (j) $7, 2, \dfrac{4}{7}, \dfrac{8}{49}, \ldots$

6. Use the series representation of sine and cosine to show that $\sin 0 = 0$ and $\cos 0 = 1$.

*7.** Improve the approximation of e obtained in Example 3 by using (a) six, (b) seven, and (c) eight terms of the series for e^x with $x = 1$.

*8.** Compute the value of $\ln 2$ by using series (four terms).

9. Use the series representation of sine and cosine to approximate $\sin 1$ and $\cos 1$. Use four terms of each series. Note that 1 is measured in radians, not degrees.

*10.** Fourier series, derived using calculus, lead to the following infinite-series representation for the irrational number π.

$$\pi = \frac{4}{1} - \frac{4}{3} + \frac{4}{5} - \frac{4}{7} + \cdots$$

Approximate π by using (a) four, (b) five, (c) six, and (d) seven terms of the series. The value of π correct to five decimal places is 3.14159.

11. Use Euler's formula to obtain the results:

$$e^{\pi i} = -1 \qquad \text{and} \qquad e^{2\pi i} = 1$$

8.2 MATHEMATICAL INDUCTION

A student has been examining sums of the form $4 + 8 + 12 + 16 + \cdots + 4n$.

When $n = 1$, $4 = 4$

When $n = 2$, $4 + 8 = 12$

When $n = 3$, $4 + 8 + 12 = 24$

When $n = 4$, $4 + 8 + 12 + 16 = 40$

Several guesses and some insight lead him to the conclusion that the sum of n terms of that form equals $2n(n + 1)$. In other words,

$$4 + 8 + 12 + 16 + \cdots + 4n = 2n(n + 1)$$

This he verifies for $n = 1, 2, 3, 4$ as follows.

When $n = 1$, $4 = 2(1)(1 + 1) = 4$

When $n = 2$, $4 + 8 = 2(2)(2 + 1) = 12$

When $n = 3$, $4 + 8 + 12 = 2(3)(3 + 1) = 24$

When $n = 4$, $4 + 8 + 12 + 16 = 2(4)(4 + 1) = 40$

But how can he *prove* that for any n, $4 + 8 + 12 + \cdots + 4n = 2n(n + 1)$? Even if he demonstrates that the equality is also true for $n = 5, 6, 7, 8, 9$, and 10, he cannot be sure that it will hold true for $n = 25$ or $n = 103$. Proof of the equality for all positive integers n is accomplished by a method called *mathematical induction*.

Proof by mathematical induction proceeds as follows. First, show that the equality is indeed true for the smallest intended value of n. Usually, this means verifying the equation for $n = 1$. Then, assume that the equality is true for $n = k$ and show this assumption assures that the equality is also true for the next n, $n = k + 1$.

The proof suggests that if the equality is true for 1, then it must work for $1 + 1$ or 2. If it works for 2, then it must work for $2 + 1$, or 3. If it holds true for 3, it must hold true for 4; and so on. Proving that if it works for k, then it must work for $k + 1$, and demonstrating that it does indeed work for 1 proves that it works for $1, 2, 3, 4, 5, \ldots$.

Imagine seven dominoes placed in a row.

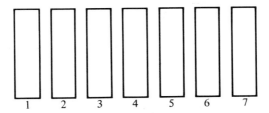

If the first domino falls to the right, then the second domino will also fall. If the second domino falls, then the third will fall. If the third falls, the fourth will fall. And so on. Observation leads to the conclusion that if domino k falls, then domino $k + 1$ must also fall. In other words, if you assume domino k falls, then you know that domino $k + 1$ will fall. So what happens when domino 1 falls? One falls, so two falls. Two falls, so three falls, ... All the dominoes fall.

Now let's use mathematical induction to prove that

$$4 + 8 + 12 + \cdots + 4n = 2n(n + 1)$$

First let $n = 1$ to see if the statement is true for the first possible value of n. When 1 is substituted for n in the equation, the equation becomes

$$4 = 2(2)$$

or

$$4 = 4$$

which is true. Next, assume that the equality is true for $n = k$. In other words, assume that

$$4 + 8 + 12 + \cdots + 4k = 2k(k + 1)$$

Finally, use the equality above to prove that the original equation is true for $n = k + 1$.

$$4 + 8 + 12 + \cdots + 4k + 4(k + 1) \overset{?}{=} 2(k + 1)(k + 2)$$

This can be done here by using the assumption that

$$4 + 8 + 12 + \cdots + 4k \quad \text{equals} \quad 2k(k + 1)$$

Therefore $2k(k + 1)$ can be substituted for $4 + 8 + 12 + \cdots + 4k$ in

$$4 + 8 + 12 + \cdots + \underbrace{4k + 4(k + 1)} \overset{?}{=} 2(k + 1)(k + 2)$$

to give

$$2k(k + 1) + 4(k + 1) \overset{?}{=} 2(k + 1)(k + 2)$$

The left side can now be manipulated into the form $2(k + 1)(k + 2)$. First factor out $(k + 1)$ from both terms on the left side.

$$(2k + 4)(k + 1) \overset{?}{=} 2(k + 1)(k + 2)$$

Next, factor 2 out of $2k + 4$.

$$2(k + 2)(k + 1) \overset{?}{=} 2(k + 1)(k + 2)$$

Finally, interchange the factors $(k + 2)$ and $(k + 1)$. The result proves the equality.

$$2(k + 1)(k + 2) = 2(k + 1)(k + 2) \quad \checkmark$$

Example 4. *Use mathematical induction to prove the following equality.*

$$\frac{1}{1 \cdot 2} + \frac{1}{2 \cdot 3} + \frac{1}{3 \cdot 4} + \cdots + \frac{1}{n(n + 1)} = \frac{n}{n + 1}$$

First, the equality is true for $n = 1$, since

$$\frac{1}{1 \cdot 2} = \frac{1}{1 + 1}$$

$$\frac{1}{2} = \frac{1}{2}$$

Now, assume that it is true for $n = k$.

$$\frac{1}{1 \cdot 2} + \frac{1}{2 \cdot 3} + \frac{1}{3 \cdot 4} + \cdots + \frac{1}{k(k + 1)} = \frac{k}{k + 1}$$

Use this result to show that the equality must be true for $n = k + 1$. That is, show that

$$\frac{1}{1 \cdot 2} + \frac{1}{2 \cdot 3} + \frac{1}{3 \cdot 4} + \cdots + \frac{1}{k(k + 1)} + \frac{1}{(k + 1)(k + 2)} = \frac{k + 1}{k + 2}$$

Use the assumption to substitute

$$\frac{k}{k+1}$$

for

$$\frac{1}{1\cdot2}+\frac{1}{2\cdot3}+\frac{1}{3\cdot4}+\cdots+\frac{1}{k(k+1)}$$

in

$$\frac{1}{1\cdot2}+\frac{1}{2\cdot3}+\frac{1}{3\cdot4}+\cdots+\frac{1}{k(k+1)}+\frac{1}{(k+1)(k+2)}\overset{?}{=}\frac{k+1}{k+2}$$

to get

$$\frac{k}{k+1}+\frac{1}{(k+1)(k+2)}\overset{?}{=}\frac{k+1}{k+2}$$

Now, find a common denominator for the fractions on the left and add them.

$$\frac{k}{(k+1)}\cdot\frac{(k+2)}{(k+2)}+\frac{1}{(k+1)(k+2)}\overset{?}{=}\frac{k+1}{k+2}$$

or

$$\frac{k(k+2)+1}{(k+1)(k+2)}\overset{?}{=}\frac{k+1}{k+2}$$

The numerator, $k(k+2)+1$, can be written as k^2+2k+1 and then factored as $(k+1)(k+1)$. The result is

$$\frac{(k+1)(k+1)}{(k+1)(k+2)}\overset{?}{=}\frac{k+1}{k+2}$$

which reduces to

$$\frac{k+1}{k+2}=\frac{k+1}{k+2}\quad\checkmark$$

Thus, the equality has been proved true for all positive integers n.

EXERCISE SET B

Prove each result by using mathematical induction.

1. $1+2+3+4+\cdots+n=\dfrac{n(n+1)}{2}$

2. $1 + 3 + 5 + \cdots + (2n - 1) = n^2$

3. $\dfrac{1}{2} + \dfrac{1}{2^2} + \dfrac{1}{2^3} + \cdots + \dfrac{1}{2^n} = 1 - \dfrac{1}{2^n}$

4. $1 + 5 + 9 + \cdots + (4n - 3) = n(2n - 1)$

5. $1^2 + 2^2 + 3^2 + \cdots + n^2 = \dfrac{n(n + 1)(2n + 1)}{6}$

6. $1 \cdot 2 + 2 \cdot 3 + 3 \cdot 4 + \cdots + n(n + 1) = \dfrac{n(n + 1)(n + 2)}{3}$

7. $1^3 + 2^3 + 3^3 + \cdots + n^3 = \dfrac{n^2(n + 1)^2}{4}$

8. $2^2 + 4^2 + 6^2 + \cdots + (2n)^2 = \dfrac{2n(n + 1)(2n + 1)}{3}$

9. $a + (a + d) + (a + 2d) + \cdots + [a + (n - 1)d] = \dfrac{n}{2}[2a + (n - 1)d]$

10. $a + ar + ar^2 + \cdots + ar^{n-1} = \dfrac{a(1 - r^n)}{1 - r}$

11. $(ab)^n = a^n b^n$

8.3 ITERATION

The concept of iteration shall be explained by means of an example. Approximations of the square root of n can be obtained by using the formula

$$x_{i+1} = \frac{1}{2}\left(x_i + \frac{n}{x_i}\right) \qquad (n > 0)$$

Supply a guess (or initial value) for $\sqrt{n}$. The guess is called x_1. Then substitute x_1 and n into the formula to get x_2, a better approximation to $\sqrt{n}$.

$$x_2 = \frac{1}{2}\left(x_1 + \frac{n}{x_1}\right)$$

Now use x_2 for x_i in the formula to produce a still better approximation, x_3.

$$x_3 = \frac{1}{2}\left(x_2 + \frac{n}{x_2}\right)$$

An approximation x_4 can be obtained by using x_3 and n in the formula. The process can be continued indefinitely. Usually a decision is made to stop once two successive approximations are "close enough" to each other. For example, if two consecutive approximations differ by less than one thousandth, that is,

$$|x_{i+1} - x_i| < .001$$

then that might be good enough.

The repetitious procedure suggested above is an example of an *iterative* process. The number obtained at any step in the process is substituted back into the formula to produce the next successive result. This iterative process for obtaining square roots is not new. It was used by the early Greeks. If you examine the formula, you'll see why it works. The right side is an average of the estimate and the number obtained by dividing the estimate into the number whose square root is sought. If the approximation x_i is too small, then n/x_i will be larger than x_i. Averaging x_i and n/x_i thus increases the size of the approximation that will be x_{i+1}. If the approximation x_i is too large, then n/x_i will be smaller than x_i. The average of x_i and n/x_i will thus produce a smaller approximation x_{i+1}.

Let's use this process to approximate $\sqrt{19}$. Let $x_1 = 4$ be the initial approximation. This means that $n = 19$ and $x_1 = 4$. So

$$x_2 = \frac{1}{2}\left(x_1 + \frac{n}{x_1}\right) = \frac{1}{2}\left(4 + \frac{19}{4}\right)$$

This yields $x_2 = 4.375$. Next, compute x_3 from

$$x_3 = \frac{1}{2}\left(x_2 + \frac{n}{x_2}\right) = \frac{1}{2}\left(4.375 + \frac{19}{4.375}\right) = 4.3589$$

The process can be continued. The square root of 19, to five decimal places, is 4.35889.

Next we show a partial flow chart of the logic of the general iterative process of approximating $\sqrt{n}$ beginning with initial value x_1.

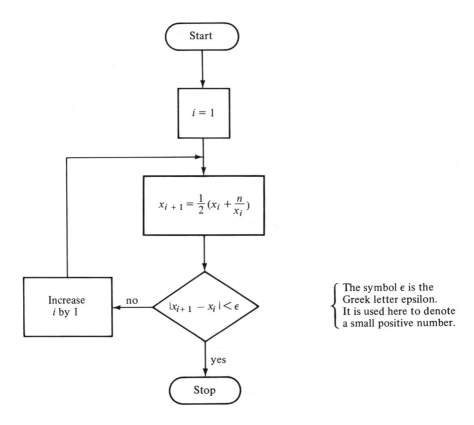

The formula used to obtain the square-root approximations is a special case of a more general application called *Newton's method.* Isaac Newton, a founder of the calculus, developed a technique for approximating zeros of functions by using the calculus. When the method is applied to $f(x) = x^2 - n$, his formula becomes the same as the early Greeks' formula. Note that the zeros of $f(x) = x^2 - n$ are the square roots of n (positive and negative). If x_1 is selected positive, then the successive approximations will approach $\sqrt{n}$. However, if the initial guess is negative, then the successive approximations will approach $-\sqrt{n}$.

EXERCISE SET C

1. Use the iterative procedure demonstrated in this section to approximate each square root. Continue until two successive approximations agree in the third place after the decimal. Use an integer approximation x_1 in each exercise.

⋆(a) $\sqrt{2}$ (b) $\sqrt{3}$ (c) $\sqrt{5}$

(d) $\sqrt{7}$ ⋆(e) $\sqrt{11}$ (f) $\sqrt{34}$

⋆(g) $\sqrt{107}$ (h) $\sqrt{151}$ (i) $\sqrt{274}$

2. Use the formula below to approximate each cube root. Use an integer approximation for x_1 and continue until $|x_{i+1} - x_i| < 0.01$.

$$x_{i+1} = \frac{1}{3}\left(2x_i + \frac{n}{x_i^2}\right)$$

⋆(a) $\sqrt[3]{10}$ (b) $\sqrt[3]{23}$ ⋆(c) $\sqrt[3]{55}$

(d) $\sqrt[3]{73}$ ⋆(e) $\sqrt[3]{100}$ (f) $\sqrt[3]{250}$

⋆Answers to starred problems are to be found at the back of the book.

chapter nine

permutations, combinations, and probability

This chapter introduces basic theory and problems of permutations and combinations. Later, that theory is applied to obtain the binomial theorem, and it is used with basic probability theory.

9.1 PERMUTATIONS AND COMBINATIONS

To begin, consider a woman who owns four different slacks (white, yellow, red, and blue) and three different sweaters (black, green, and blue). In how many different ways can she select both slacks and sweater to wear? She can begin by selecting any of the four slacks. Then, for any slacks selected there are any of three sweaters to go with them. This means that there are $4 \cdot 3$, or 12, different selections altogether. In general, if one event can occur in M ways *and* an event following it can occur in N ways, then both events can occur in $M \cdot N$ ways. If the woman were selecting slacks or a sweater—one or the other (but not both)— then she would have $4 + 3$, or 7, choices. In general, if one event can occur in M ways *or* another event can occur in N ways, then one or the other can occur in $M + N$ ways. Thus, in these settings, "and" leads to multiplication; "or" yields addition. If more than two events are occurring, then the sum or product will contain more than two terms. A and B and C and D leads to $A \cdot B \cdot C \cdot D$. Similarly, A or B or C or D leads to $A + B + C + D$.

A *permutation* is an order or arrangement. The letters a, b, c, d, and e can be arranged in 120 different ways. In other words, there are 120 permutations of the letters a, b, c, d, and e. To see this without actually listing all of them, consider five slots, each to be filled with one letter. Any of the five letters can be placed in the first slot.

$$\underline{5}\ \underline{\ }\ \underline{\ }\ \underline{\ }\ \underline{\ }$$

Then there will be only four letters available for placement in the next slot.

$$\underline{5}\ \underline{4}\ \underline{\ }\ \underline{\ }\ \underline{\ }$$

Continuing this way, there are successively 3, 2, and 1 letters for the remaining slots.

$$\underline{5}\ \underline{4}\ \underline{3}\ \underline{2}\ \underline{1}$$

The placing of five letters can be considered five events, each occurring one right after the other. That is, in succession we place any of

$$5 \quad \text{and} \quad 4 \quad \text{and} \quad 3 \quad \text{and} \quad 2 \quad \text{and} \quad 1$$

letters. So the number of ways this can be done is $5 \cdot 4 \cdot 3 \cdot 2 \cdot 1$, or 5! The number 5! simplifies to 120 when multiplied out. Note for future reference that 0! is defined to be 1; that is, $0! = 1$. This follows from observing $n! = n(n - 1)!$ with $n = 1$.

Example 1. *A club has 15 members, from whom will be elected a president, vice-president, secretary, and treasurer. In how many ways can this selection be made?*

There are 15 possibilities for president. After he is elected, there are 14 members left who could be vice-president. After that selection, there will be 13 persons available for secretary, and finally 12 for treasurer. The club is electing a president *and* vice-president *and* secretary *and* treasurer. Thus, the selection can occur in $15 \cdot 14 \cdot 13 \cdot 12$, or 32,760 ways.

If some of the elements used in permutations are the same, then there will be fewer distinguishable permutations than if all are different. For example, the set $\{a, b, c, d\}$ has 4!, or 24, permutations. They are:

abcd	*bacd*	*cabd*	*dabc*
abdc	*badc*	*cadb*	*dacb*
acbd	*bcad*	*cbad*	*dbac*
acdb	*bcda*	*cbda*	*dbca*
adbc	*bdac*	*cdab*	*dcab*
adcb	*bdca*	*cdba*	*dcba*

If, instead, three of the letters are the same, as with $\{a, b, b, b\}$, then there are fewer (distinguishable) permutations. In fact, there are only four permutations.

abbb

babb

bbab

bbba

The reason there are fewer is because there are no longer four different letters. If the three *b*'s were different, then there would be 3! times as many permutations. Three different letters would contribute 3! permutations to the product.

$a, \underline{b}, \underline{b}, \underline{b}$ 4 permutations

$a, \underline{b}, \underline{c}, \underline{d}$ 4·3! permutations

Note that 4·3! is the same as 4!, since $4 \cdot 3! = 4 \cdot 3 \cdot 2 \cdot 1 = 4!$.

Working backward, you should see that the number of permutations of $\{a, b, b, b\}$, namely four, could have been computed by dividing 4! by 3!, that is, by dividing out the effect of the three *b*'s being the same. In other words, divide out the 3! orders which are not present because the *b*'s are the same. Thus, the number of permutations of $\{a, b, b, b\}$ is $4!/3! = 4$.

Example 2. *How many permutations are possible using all the letters of the word ALGEBRA?*

If all seven letters were different, then there would be 7! permutations. However, there are two A's, so divide out the 2! orders they would contribute to 7! if they were different. The number of permutations of the letters of the word *ALGEBRA* is thus 7!/2!, or 2520.

Example 3. *A man has 10 poker chips; 5 red, 3 blue, 2 white. If he arranges them in a pile, one on top of the other, how many distinguishable piles (orders) are possible?*

If all the chips were different colors, then there would be 10! arrangements possible. Since the 5 red ones are the same, the 3 blue ones are the same, and the 2 white ones are the same, divide out the 5!, 3!, and 2! orders that are built into the 10!. The number of different arrangements is

$$\frac{10!}{5!\,3!\,2!} \quad \checkmark$$

If the five letters of the set $\{a, b, c, d, e\}$ are taken two at a time for different arrangements of two each, then there will be $5 \cdot 4$, or 20, arrangements. This also happens to be $5!/3!$ arrangements. They are

ab	ad	bc	be	ce
ba	da	cb	eb	ec
ac	ae	bd	cd	de
ca	ea	db	dc	ed

Suppose now that we change our intention and insist that we don't care about different orders; we only want to count different *combinations* of letters. In other words, *ab* and *ba* are different *permutations* (or arrangements), but they are the same *combination* of letters. Both are *a* and *b*. To get combinations, divide out the order built into the $5!/3!$ permutations. In each case there are two letters used, so divide by 2! to eliminate the orders. Thus, if

$$\text{Permutations} = \frac{5!}{3!} = 20$$

then

$$\text{Combinations} = \frac{\frac{5!}{3!}}{2!} = \frac{5!}{3!\,2!} = 10$$

To convert from permutations to combinations, divide by the factorial of the number of elements selected for each combination. This will divide out the order.

Example 4. *From the set $\{b, c, p, r, t, x, y, z\}$ elements are selected five at a time. Determine (a) the number of permutations; (b) the number of combinations that are possible.*

(a) The number of permutations is

$$8 \cdot 7 \cdot 6 \cdot 5 \cdot 4 \quad \text{or} \quad \frac{8!}{3!} \quad \checkmark$$

(b) The number of combinations is determined by dividing the number of permutations by 5!, since there are five elements in each combination. The number of combinations is

$$\frac{8!}{3!\,5!} \quad \checkmark$$

To avoid relying on permutations for every problem involving combinations, it is desirable to have a general form for expressing combinations. The notation $\binom{5}{2}$ is used to mean the number of combinations of 5 things selected 2 at a time. Similarly, $\binom{8}{3}$ is the number of combinations of 8 things taken 3 at a time. For computation,

$$\binom{5}{2} = \frac{5!}{2!\,3!} = \frac{5\cdot 4\cdot 3!}{2\cdot 3!} = 10$$

$$\binom{8}{3} = \frac{8!}{3!\,5!} = \frac{8\cdot 7\cdot 6\cdot 5!}{3\cdot 2\cdot 1\cdot 5!} = 56$$

$$\binom{n}{r} = \frac{n!}{r!\,(n-r)!}$$

$$\binom{48}{2} = \frac{48!}{2!\,46!} = \frac{48\cdot 47\cdot 46!}{2\cdot 46!} = 1128$$

Example 5. *In how many ways can a committee of 4 people be selected from a group of 7 people?*

Committees can be selected in $\binom{7}{4}$ ways. And

$$\binom{7}{4} = \frac{7!}{4!\,3!} = \frac{7\cdot 6\cdot 5\cdot 4!}{3\cdot 2\cdot 1\cdot 4!} = 35$$

Note that the committees selected are combinations. Each committee consists of three people, and it is the same committee regardless of the order in which the members are selected.

Example 6. *How many committees of 3 men and 4 women can be selected from 10 men and 9 women?*

Three men can be selected from 10 men in $\binom{10}{3}$ ways. Similarly, 4 women can be selected from 9 women in $\binom{9}{4}$ ways. These are two separate events: choosing men and choosing women. Thus, the number of committees that can be formed is the product of $\binom{10}{3}$ and $\binom{9}{4}$, or

$$\binom{10}{3}\binom{9}{4} \quad \checkmark$$

Example 7. *In how many ways can a class of* 24 *students be divided into three groups—one of* 10 *students, one of* 5 *students, and one of* 9 *students?*

At the beginning there are 24 students. We can select the first group of 10 in $\binom{24}{10}$ ways. This leaves 14 students for the other groups. Thus, the next group, say 5 students, can be selected in $\binom{14}{5}$ ways. This leaves 9 students for the last group, so there are $\binom{9}{9}$ ways of selecting it. The selection of the three groups can be considered three separate events, one following the other; so that the total number of different ways in which the class can be divided is the product of the three combinations above, $\binom{24}{10}\cdot\binom{14}{5}\cdot\binom{9}{9}$ This product can be reduced somewhat.

$$\binom{24}{10}\binom{14}{5}\binom{9}{9} = \frac{24!}{10!\,14!} \cdot \frac{14!}{5!\,9!} \cdot \frac{9!}{9!\,0!} = \frac{24!}{10!\,5!\,9!} \quad \checkmark$$

EXERCISE SET A

*1. In how many different ways can the letters of the word *complex* be arranged?

*2. In how many different ways can a president and secretary be chosen from a group of 10 people?

3. How many permutations are possible using for each permutation four of the letters b, c, d, e, f, and g?

4. Five students walk into a classroom which has 10 seats in a row. In how many different (distinguishable) ways can the students sit in those seats?

*5. A test has six true–false questions. It also has four multiple-choice questions with five selections possible for each question. In how many different ways can the test questions be answered?

*6. How many different license plates can be made in the form of two letters followed by three digits? Letters and digits can be repeated on any plate.

7. A student has three different math books and five different art books. In how many ways can she arrange them on a shelf, if the three math books must come first (left) before the art books are placed on the shelf?

*8. How many different arrangements of the word *mathematics* are possible?

*Answers to starred problems are to be found at the back of the book.

★9. Art has 20 brand new coins. Seven are silver, 9 are bronze, and 4 are gold. How many different-looking stacks of 20 can he make?

10. Doris dips her spoon into her alphabet soup and obtains the following: three B's, two X's, one N, and one S. She then takes the letters and arranges them in a straight line. How many different arrangements are possible?

★11. Compute and simplify each.

(a) $\dbinom{6}{2}$

(b) $\dbinom{10}{2}$

(c) $\dbinom{32}{1}$

(d) $\dbinom{11}{8}$

(e) $\dbinom{15}{15}$

(f) $\dbinom{14}{0}$

12. Show that each relationship is true.

(a) $\dbinom{n}{1} = n$

(b) $\dbinom{n}{n} = 1$

(c) $\dbinom{n}{x} = \dbinom{n}{n-x}$

★13. In how many ways can a committee of 5 people be selected from a group of 14 people?

14. In how many ways can a committee of 2 students and 3 faculty members be selected from 100 students and 50 faculty?

★15. In how many ways can two committees, one of size 6 and one of size 17, be chosen from a group of 23 people? No person can be on both committees.

16. In how many ways can a group of 16 people be divided into two groups of 8 people each?

★17. From a club of 18 members, one person is elected chairman of a committee to be formed. He or she then selects the other 3 members, who will compose the committee of 4. In how many ways can this election-selection process occur?

★18. How many different committees of 3 or 4 members each can be selected from 9 men and 6 women, if all committees must have at least 1 man and 1 woman?

19. In how many ways can 4 or fewer birds be selected from a nest of 11 birds?

20. Simplify each expression completely.

(a) $\dfrac{(n+1)!}{n+1}$

(b) $\dfrac{n!}{(n+1)!}$

(c) $\dfrac{(n+2)!}{n!}$

(d) $\dfrac{\dfrac{x^{n+1}}{(n+1)!}}{\dfrac{x^n}{n!}}$

9.2 SUMMATION NOTATION

A capital sigma, Σ, is used frequently in mathematics to denote a sum. Here is an example of such use.

$$1 + 2 + 3 + 4 + 5 + 6 = \sum_{k=1}^{6} k$$

Note the compact summation form on the right side. It indicates the sum of numbers of the form k. Below the sigma, $k = 1$ specifies that k begins at 1. The 6 on top of the sigma indicates that 6 is the last value that k takes on. Thus, k begins at 1 and *counts* 1, 2, 3, 4, 5, 6. Here is another example:

$$\begin{aligned}
\sum_{k=2}^{5} k^2 &= 2^2 + 3^2 + 4^2 + 5^2 \\
&= 4 + 9 + 16 + 25 \\
&= 54 \;\checkmark
\end{aligned}$$

The count is always by 1's, that is, by consecutive integers. The counter or index letter need not be k. Often i, j, and n are used. Any letter can be used.

$$\begin{aligned}
\sum_{i=1}^{5} (i + 1) &= (1 + 1) + (2 + 1) + (3 + 1) + (4 + 1) + (5 + 1) \\
&= 2 \qquad + 3 \qquad + 4 \qquad + 5 \qquad + 6 \\
&= 20 \;\checkmark
\end{aligned}$$

$$\begin{aligned}
\sum_{j=0}^{3} \frac{1}{j+1} &= \frac{1}{0+1} + \frac{1}{1+1} + \frac{1}{2+1} + \frac{1}{3+1} \\
&= 1 \quad\;\; + \frac{1}{2} \;\;\cdot + \frac{1}{3} \quad + \frac{1}{4} \\
&= \frac{25}{12} \;\checkmark
\end{aligned}$$

$$\begin{aligned}
\sum_{n=-1}^{4} 2n &= 2(-1) + 2(0) + 2(1) + 2(2) + 2(3) + 2(4) \\
&= -2 \quad + 0 \quad + 2 \quad + 4 \quad + 6 \quad + 8 \\
&= 18 \;\checkmark
\end{aligned}$$

9.3 BINOMIAL THEOREM

This section presents a technique for expanding powers of binomials, two-term expressions. It is called the *binomial theorem*. Observe the results of raising $(a + b)$ to whole powers.

$$(a + b)^0 = 1$$
$$(a + b)^1 = a + b$$
$$(a + b)^2 = a^2 + 2ab + b^2$$
$$(a + b)^3 = a^3 + 3a^2b + 3ab^2 + b^3$$
$$(a + b)^4 = a^4 + 4a^3b + 6a^2b^2 + 4ab^3 + b^4$$
$$(a + b)^5 = a^5 + 5a^4b + 10a^3b^2 + 10a^2b^3 + 5ab^4 + b^5$$

In each case the terms are written in order of decreasing powers of a (and increasing powers of b). In each term, the sum of the exponents of a and b equals the power to which the binomial is raised. For example, in $(a + b)^5$, the terms are

Terms	Powers of a and b
$1a^5b^0$	$5 + 0 = 5$
$5a^4b^1$	$4 + 1 = 5$
$10a^3b^2$	$3 + 2 = 5$
$10a^2b^3$	$2 + 3 = 5$
$5a^1b^4$	$1 + 4 = 5$
$1a^0b^5$	$0 + 5 = 5$

The coefficients of each term form a pattern, which was first recognized in the 1600s by Blaise Pascal, a French mathematician. The pattern is easy to see in *Pascal's triangle*.

$$
\begin{array}{ccccccccc}
& & & & 1 & & & & \\
& & & 1 & & 1 & & & \\
& & 1 & & 2 & & 1 & & \\
& 1 & & 3 & & 3 & & 1 & \\
1 & & 4 & & 6 & & 4 & & 1
\end{array}
$$

and so on

Each row of Pascal's triangle begins and ends with 1. Any other term in any row can be obtained by adding the two terms above its position. For example, the 4 is obtained by adding the 1 and 3 above its position. Similarly, 6 comes from $3 + 3$. Generation of the next row of coefficients, for $(a + b)^5$, is shown next.

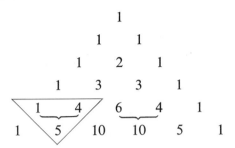

Note also that the number represented by all the digits in each of the first five rows that compose the triangle is a power of 11. The first row is 11^0, the next 11^1, then 11^2, 11^3, and 11^4. However, $11^5 \neq 15101051$.

Example 8. *Expand* $(x + 2y)^4$.

$$(x + 2y)^4 = [(x) + (2y)]^4$$

The expansion contains powers of x times powers of $2y$, with appropriate coefficients determined by Pascal's triangle. Those coefficients are 1, 4, 6, 4, and 1. Thus,

$$[(x) + (2y)]^4 = 1(x)^4(2y)^0 + 4(x)^3(2y)^1 + 6(x)^2(2y)^2$$
$$+ 4(x)^1(2y)^3 + 1(x)^0(2y)^4$$
$$= 1 \cdot x^4 \cdot 1 + 4 \cdot x^3 \cdot 2y + 6 \cdot x^2 \cdot 4y^2 + 4 \cdot x \cdot 8y^3 + 1 \cdot 1 \cdot 16y^4$$
$$= x^4 + 8x^3y + 24x^2y^2 + 32xy^3 + 16y^4 \quad \checkmark$$

Example 9. *Expand* $(4m - n)^3$.

The binomial must be written as the sum $(+)$ of two terms. Thus,

$$(4m - n)^3 = [(4m) + (-n)]^3$$

From Pascal's triangle, the coefficients will be 1, 3, 3, and 1.

$$[(4m) + (-n)]^3 = 1(4m)^3(-n)^0 + 3(4m)^2(-n)^1 + 3(4m)^1(-n)^2$$
$$+ 1(4m)^0(-n)^3$$
$$= 1 \cdot 64m^3 \cdot 1 + 3 \cdot 16m^2 \cdot -n + 3 \cdot 4m \cdot n^2 + 1 \cdot 1 \cdot -n^3$$
$$= 64m^3 - 48m^2n + 12mn^2 - n^3 \quad \checkmark$$

Although Pascal's triangle is useful for expanding binomials, it does not provide a general form for $(a + b)^n$, nor is it practical for large powers n. Let us study $(a + b)^4$ in an attempt to arrive at a general formula for $(a + b)^n$.

$$(a + b)^4 = (a + b)(a + b)(a + b)(a + b)$$

To actually multiply this out, it is necessary to take one term (a or b) from the first $a + b$ factor, times one term (a or b) from the second $a + b$ factor, times one term (a or b) from the third $a + b$ factor, times one term (a or b) from the fourth $a + b$ factor. And we must take all possible combinations in order to assure that we obtain the proper number of a^4, a^3b, a^2b^2, ab^3, and b^4 terms. Terms of the form a^4 are obtained from $(a + b)(a + b)(a + b)(a + b)$ by taking an a from each $a + b$ factor. In other words, from the four a's available, take all four. And there are $\binom{4}{4}$ ways of selecting four things out of four. You can also reason that you want to select no b's from the four b's available. This, too, will produce a^4. So the coefficient of a^4 can be called $\binom{4}{4}$ or $\binom{4}{0}$. They are equal, so it doesn't matter which is used. To fit the form we will want later, $\binom{4}{0}$ will be used.

To obtain a^3b or a^3b^1, you must select one b from the four available b's. The other three factors selected are a's. This can be done in $\binom{4}{1}$ ways.

$$(a + \underline{b})(\underline{a} + b)(\underline{a} + b)(\underline{a} + b)$$

or

$$(\underline{a} + b)(a + \underline{b})(\underline{a} + b)(\underline{a} + b)$$

or

$$(\underline{a} + b)(\underline{a} + b)(a + \underline{b})(\underline{a} + b)$$

or

$$(\underline{a} + b)(\underline{a} + b)(\underline{a} + b)(a + \underline{b})$$

The coefficient of a^3b is then $\binom{4}{1}$. Similarly, the coefficient of a^2b^2 is $\binom{4}{2}$; the coefficient of ab^3 is $\binom{4}{3}$; and the coefficient of b^4 is $\binom{4}{4}$. Thus,

$$(a + b)^4 = \binom{4}{0}a^4b^0 + \binom{4}{1}a^3b^1 + \binom{4}{2}a^2b^2 + \binom{4}{3}a^1b^3 + \binom{4}{4}a^0b^4$$

The top number in each binomial coefficient is 4, the power to which $(a + b)$ is being raised. The lower number in the coefficient is the same as the power of b, and it begins at zero and counts up to 4, the power of

$(a + b)$. The sum of the powers of a and b in any term equals 4, the power of $(a + b)$.

The *binominal theorem* can be stated for $(a + b)^n$ by using combination notation.

$$(a + b)^n = \binom{n}{0}a^n b^0 + \binom{n}{1}a^{n-1}b^1 + \binom{n}{2}a^{n-2}b^2 + \cdots$$
$$+ \binom{n}{n-1}a^1 b^{n-1} + \binom{n}{n}a^0 b^n$$

Summation notation can be used to condense this form to

$$(a + b)^n = \sum_{i=0}^{n}\binom{n}{i}a^{n-i}b^i$$

EXERCISE SET B

1. Compute the value of each expression and simplify completely.

★(a) $\sum_{i=0}^{9} i^2$

★(b) $\sum_{j=2}^{16}(j - 1)$

(c) $\sum_{n=1}^{7}(2n + 1)$

(d) $\sum_{n=0}^{5}\frac{n}{n + 1}$

★(e) $\sum_{k=0}^{6} 2^k$

(f) $\sum_{j=-2}^{5} j(j + 2)$

★(g) $\sum_{i=0}^{4}\binom{4}{i}$

(h) $\sum_{n=2}^{6}\binom{n}{n - 1}$

2. Expand each binomial by using the binomial theorem. Simplify completely.

★(a) $(c + d)^5$ (b) $(x + y)^4$
(c) $(x + y)^6$ ★(d) $(a - b)^3$
(e) $(p - q)^7$ ★(f) $(2x + y)^5$
(g) $(x - 2y)^6$ ★(h) $(2x - 3y)^5$
(i) $(x + y)^8$ (j) $(x - y)^9$
★(k) $(1.01)^6$ (l) $(0.99)^5$
★(m) $(1 - x^2)^7$

★**3.** Find the term of $(3x^2 + y)^{20}$ which contains x^4 when simplified.

★Answers to starred problems are to be found at the back of the book.

4. Find the term of $(c + 2d^2)^{10}$ which contains d^6 when simplified.

5. Find the first four terms of $(x + \Delta x)^n$.

6. Rewrite the series for e^x, $\ln x$, $\sin x$, and $\cos x$ (from Section 8.1) using $\sum$ notation. The symbol ∞ is used here as the upper limit on the sigma.

9.4 PROBABILITY

You have undoubtedly played a variety of games in which dice or cards were used. Perhaps you have tried your luck at a raffle. State lotteries are becoming increasingly popular as a means of raising revenue. All these settings suggest situations in which probability can be applied. Although the outcome cannot be predicted each time, the chances of any particular outcome occurring can be determined methodically. Elementary probability is presented in this section. Many of the examples and exercises use dice, cards, coins, and other gambling-oriented settings. These are, of course, the original applications of probability. Other problems suggest applications in other areas.

The probability that an event E will occur is denoted $P(E)$ and defined as

$$P(E) = \frac{n(E)}{T}$$

where $n(E)$ is the number of favorable outcomes, that is, the number of ways in which event E can occur. T is the total number of possible outcomes. All outcomes must be equally likely to occur.

Example 10. *A coin is flipped. What is the probability that it will land tails?*

The possible outcomes are heads or tails. Here E is the event that the coin comes up tails. Thus,

$$P(E) = \frac{n(E)}{T} = \frac{1}{2} \quad \checkmark$$

The probability that the coin lands tails is $1/2$. Of course, we have assumed that the outcomes heads and tails are equally likely; that is, the coin is just as likely to land heads as tails.

Example 11. *A single die is rolled. What is the probability that it will come up (a)* 1, 2, 3, 4, 5, or 6; *(b)* 7?

(a) Here E is the event that the die will come up 1, 2, 3, 4, 5, or 6. This means that $n(E) = 6$. Also, the only ways a die can come up are 1, 2, 3, 4, 5, or 6. This means that $T = 6$, too. Thus, $P(E) = 1$:

$$P(E) = \frac{n(E)}{T} = \frac{6}{6} = 1 \quad \checkmark$$

(b) Here E is the event that a die comes up 7. This cannot happen, so $n(E) = 0$. T is still 6, as before. This leads to $P(E) = 0$, as

$$P(E) = \frac{n(E)}{T} = \frac{0}{6} = 0 \quad \checkmark$$

From this example, the following conclusions can be drawn:

$$P(\text{certain event}) = 1$$
$$P(\text{impossible event}) = 0$$

This means that values of probabilities will be between 0 and 1 inclusive.

$$\boxed{0 \leq P(E) \leq 1}$$

Example 12. *A pair of dice is rolled. What is the probability that the sum of the two sides facing upward is 8?*

Since one die (the first) can come up any of six ways and the other die (the second) can also come up any of six ways, there are $6 \cdot 6$, or 36, possible outcomes. Although it is not necessary to list all possible outcomes, it is done below to convince any doubters. The notation is (first die, second die).

$$
\begin{array}{cccccc}
(1, 1) & (1, 2) & (1, 3) & (1, 4) & (1, 5) & (1, 6) \\
(2, 1) & (2, 2) & (2, 3) & (2, 4) & (2, 5) & (2, 6) \\
(3, 1) & (3, 2) & (3, 3) & (3, 4) & (3, 5) & (3, 6) \\
(4, 1) & (4, 2) & (4, 3) & (4, 4) & (4, 5) & (4, 6) \\
(5, 1) & (5, 2) & (5, 3) & (5, 4) & (5, 5) & (5, 6) \\
(6, 1) & (6, 2) & (6, 3) & (6, 4) & (6, 5) & (6, 6)
\end{array}
$$

So $T = 36$. The outcomes that produce a sum of 8 are (2, 6), (6, 2), (3, 5), (5, 3), and (4, 4). So $n(E) = 5$. Thus,

$$P(\text{obtain 8 on two dice}) = P(E) = \frac{n(E)}{T} = \frac{5}{36} \quad \checkmark$$

Example 13. *A die is thrown and a coin is flipped. What is the probability of getting an even number on the die and tails on the coin?*

The possible outcomes are: (1, H), (1, T), (2, H), (2, T), (3, H), (3, T), (4, H), (4, T), (5, H), (5, T), (6, H), (6, T), where the first coordinate is the outcome of the die and the second coordinate is the outcome of the coin. If E is the event that an even number is obtained on the die and tails is obtained on the coin, then $n(E) = 3$. Here are the three outcomes: (2, T), (4, T), (6, T). The total number of outcomes, displayed above, is 12. Thus,

$$P(E) = \frac{n(E)}{T} = \frac{3}{12} = \frac{1}{4} \quad \checkmark$$

Note that there are 12 possible outcomes. Since the die can come up any of 6 ways and the coin can land any of 2 ways, there are $6 \cdot 2 = 12$ possible outcomes.

$$n(\text{die } and \text{ coin}) = n(\text{die}) \cdot n(\text{coin})$$
$$= 6 \cdot 2$$
$$= 12$$

This can be extended to probabilities as

$$P(\text{event die is even } and \text{ event coin is T}) = \frac{n(\text{die even } and \text{ coin T})}{n(\text{die anything } and \text{ coin anything})}$$

$$= \frac{n(\text{die even}) \cdot n(\text{coin T})}{n(\text{die any}) \cdot n(\text{coin any})}$$

$$= \frac{n(\text{die even})}{n(\text{die any})} \cdot \frac{n(\text{coin T})}{n(\text{coin any})}$$

$$= \frac{3}{6} \cdot \frac{1}{2} = \frac{3}{12} \quad \text{or} \quad \frac{1}{4}$$

What has been shown above is an example of the major result that for two events E and F,

$$\boxed{P(E \text{ and } F) = P(E) \cdot P(F)}$$

Warning: *If $P(F)$ is influenced by the consideration of $P(E)$, then the events E and F are* dependent. *In that case, extreme care must be used in computing $P(F)$ after determining $P(E)$. If the events E and F are not dependent, then they are* independent.

In the last example, the events "getting an even number on a die" and "getting tails on a coin" are independent. Whether or not an even number appears on the die in no way affects the outcome of the coin flip. An example of dependent events will be given in Example 16.

Example 14. *A card is drawn from a standard deck of 52 cards. Then it is replaced and another card is drawn from the deck. What is the probability that both cards drawn are hearts?*

Let

$E =$ event hearts is obtained on the first card

$F =$ event hearts is obtained on the second card

Then we seek $P(E$ and $F)$.

$$P(E \text{ and } F) = P(E) \cdot P(F)$$
$$= \frac{n(E)}{T} \cdot \frac{n(F)}{T}$$
$$= \frac{13}{52} \cdot \frac{13}{52} \quad \text{or} \quad \frac{1}{16} \quad \checkmark \quad (13 \text{ of the 52 cards are hearts})$$

The experiment of Example 14 was performed *with replacement*; that is, the first card was replaced before the second card was selected.

Example 15. *There are two boxes on a table. One contains 4 red balls and 3 green balls. The other contains 5 blue balls and 5 yellow balls. A box is selected at random and 1 ball is drawn from it. What is the probability that a green ball is drawn?*

In order to pick a green ball, you must first select the right box. So consider

$E =$ event box containing green balls is selected

$F =$ event green ball is selected from that box

Then

$$P(\text{event green ball is selected}) = P(E \text{ and } F)$$
$$= P(E) \cdot P(F)$$
$$= \frac{1}{2} \cdot \frac{3}{7} \quad \text{or} \quad \frac{3}{14} \quad \checkmark$$

It has been assumed that it is just as likely to select either box. Hence, each box is selected with probability $\frac{1}{2}$. The box with green balls contains 3 green balls and 4 red balls—7 altogether. So the probability of picking a green ball from this box is $\frac{3}{7}$.

Example 16. *A card is drawn from a deck of 52. Then another card is drawn from the same deck without replacing the first card (that is, without replacement). What is the probability that both cards drawn are hearts?*

Let

E = event hearts is selected on the first draw

F = event hearts is selected on the second draw

Then,

$$P(E) = \frac{13}{52}$$

since there are 52 cards and 13 of them are hearts.

$$P(F) = \frac{12}{51}$$

since there are now only 51 cards (one was removed by the first draw) and 12 of them are hearts (since the card drawn first must be hearts if we are to get hearts in two draws. Thus,

$$P(E \text{ and } F) = P(E) \cdot P(F)$$

$$= \frac{13}{52} \cdot \frac{12}{51} \quad \text{or} \quad \frac{1}{17} \quad \checkmark$$

Note that the two events E and F are dependent. Consideration of the occurrence of event E affects the computation of $P(F)$.

Example 17. *Two cards are drawn from a deck of 52 without replacement. What is the probability that the first is a heart and the second is a spade?*

Let

E = event first card selected is a heart

F = event second card selected is a spade

Then

$$P(E \text{ and } F) = P(E) \cdot P(F)$$

$$= \frac{13}{52} \cdot \frac{13}{51} \quad \text{or} \quad \frac{13}{204} \quad \checkmark$$

There are only 51 cards left for the second draw, and all 13 spades are still in the deck.

Selecting cards from a deck without replacement is like selecting committees. For example, 2 cards can be selected in $\binom{52}{2}$ ways. Two hearts can be selected from 13 hearts in $\binom{13}{2}$ ways. So if E is the event that 2 hearts are drawn in two draws without replacement from a 52-card deck, then

$$P(E) = \frac{n(E)}{T} = \frac{\binom{13}{2}}{\binom{52}{2}}$$

which reduces to the $\frac{13}{52} \cdot \frac{12}{51}$ form of Example 16.

Example 18. *A bag contains 5 blue marbles and 8 red marbles. Four are selected without replacement. What is the probability that 2 are blue and 2 are red?*

The number of ways in which 2 blue marbles can be selected is $\binom{5}{2}$, since there are 5 blue marbles to choose from. Similarly, the 2 red marbles can be selected in $\binom{8}{2}$ ways. Thus, 2 blue marbles *and* 2 red marbles can be selected in $\binom{5}{2}\binom{8}{2}$ ways. The total number of ways in which 4 marbles can be selected from 13 marbles (5 blue, 8 red) is $\binom{13}{4}$. Thus, if E is the event that 2 blue marbles and 2 red marbles are selected, then

$$P(E) = \frac{n(E)}{T} = \frac{\binom{5}{2}\binom{8}{2}}{\binom{13}{4}}$$

This probability can be determined without using combinations (as above), but a problem of order arises, as shown next.

Let us consider the problem as that of selecting four marbles, one right after another. The probability of getting a blue marble first is $\frac{5}{13}$, since there are 13 marbles, 5 of which are blue. If a blue marble is selected, then the probability is $\frac{4}{12}$ that a blue marble will be selected next, since only 12 marbles are left and only 4 of those are blue. The probability of getting a red marble on the third pick is $\frac{8}{11}$. The probability of getting a red marble on the fourth pick is $\frac{7}{10}$. Thus, it may appear that

$$P(E) = \frac{5}{13} \cdot \frac{4}{12} \cdot \frac{8}{11} \cdot \frac{7}{10}$$

But this is wrong! Actually, it is just incomplete. The product above represents the probability of selecting the sequence: blue, blue, red, red.

There are five other ways to obtain 2 blue marbles and 2 red marbles:

blue, red, blue, red $\quad\left(\text{with probability}\quad \frac{5}{13}\cdot\frac{8}{12}\cdot\frac{4}{11}\cdot\frac{7}{10}\right)$

blue, red, red, blue $\quad\left(\text{with probability}\quad \frac{5}{13}\cdot\frac{8}{12}\cdot\frac{7}{11}\cdot\frac{4}{10}\right)$

red, blue, blue, red $\quad\left(\text{with probability}\quad \frac{8}{13}\cdot\frac{5}{12}\cdot\frac{4}{11}\cdot\frac{7}{10}\right)$

red, blue, red, blue $\quad\left(\text{with probability}\quad \frac{8}{13}\cdot\frac{5}{12}\cdot\frac{7}{11}\cdot\frac{4}{10}\right)$

red, red, blue, blue $\quad\left(\text{with probability}\quad \frac{8}{13}\cdot\frac{7}{12}\cdot\frac{5}{11}\cdot\frac{4}{10}\right)$

Each of the *six* sequences is obtained with the same probability,

$$\frac{5}{13}\cdot\frac{4}{12}\cdot\frac{8}{11}\cdot\frac{7}{10}$$

Thus,

$$P(E) = 6\cdot\frac{5}{13}\cdot\frac{4}{12}\cdot\frac{8}{11}\cdot\frac{7}{10}$$

The six orders could have been determined without listing each one. It is

$$\frac{4!}{2!\,2!}\quad\text{or}\quad\binom{4}{2}$$

If the problem involved 3 blue and 4 red (seven altogether), then the number of orders would be

$$\frac{7!}{3!\,4!}\quad\text{or}\quad\binom{7}{3}\quad\text{or}\quad\binom{7}{4}$$

Example 19. *A bag contains 5 blue marbles and 8 red marbles. Four of them are selected with replacement. What is the probability that 2 are blue and 2 are red?*

This problem is similar to Example 18. However, because of replacement there will always be 13 marbles in the bag; there will always be 5 blue ones and 8 red ones to choose from. Thus,

$$P(E) = \binom{4}{2}\cdot\frac{5}{13}\cdot\frac{5}{13}\cdot\frac{8}{13}\cdot\frac{8}{13}\quad\checkmark$$

Example 20. *A box contains 4 red marbles, 8 yellow marbles, 5 blue marbles, and 6 green marbles. Seven marbles are drawn without replacement. What is the probability that 2 are red, 3 are blue, and 2 are green?*

The 2 red, 3 blue, and 2 green marbles can be selected in several orders. In fact, there are $\frac{7!}{2!\,3!\,2!}$ orders. If all were different colors, there would be 7! orders. Here the 2!, 3!, and 2! divide out the order that would be present if they were all different colors.

The probability of getting a particular order, say *RRBBBGG*, is

$$\frac{4}{23} \cdot \frac{3}{22} \cdot \frac{5}{21} \cdot \frac{4}{20} \cdot \frac{3}{19} \cdot \frac{6}{18} \cdot \frac{5}{17}$$

So if *E* is the event that 2 red, 3 blue, and 2 green marbles are drawn, then

$$P(E) = \frac{7!}{2!\,3!\,2!} \cdot \frac{4}{23} \cdot \frac{3}{22} \cdot \frac{5}{21} \cdot \frac{4}{20} \cdot \frac{3}{19} \cdot \frac{6}{18} \cdot \frac{5}{17} \quad \checkmark$$

Problems involving the probability of one event *or* another event are presented next.

Example 21. *Two dice are rolled. What is the probability that the sum of the upward faces is 7 or 8?*

There are six outcomes possible for each of the two dice, so there are $6 \cdot 6 = 36$ total outcomes of the two dice together. Six of these have sums of 7 (1-6, 6-1, 2-5, 5-2, 3-4, 4-3) and 5 of the 36 have sums of 8 (2-6, 6-2, 3-5, 5-3, 4-4). Thus, there are 11 favorable outcomes and 36 total outcomes. So if *E* is the event that the outcome is a 7 or an 8, then

$$P(E) = \frac{11}{36} \quad \checkmark$$

On the other hand, suppose that we let *A* be the event the outcome is a 7 and let *B* be the event the outcome is an 8. Note that

$$P(A) = \frac{6}{36}$$

$$P(B) = \frac{5}{36}$$

and

$$P(E) = P(A \text{ or } B) = P(A) + P(B)$$

$$= \frac{6}{36} + \frac{5}{36}$$

$$= \frac{11}{36} \quad \checkmark$$

In general, if A and B are two events, then $P(A \text{ or } B) = P(A) + P(B)$ if A and B are *mutually exclusive events*, that is, if A and B cannot both occur simultaneously. Occurrence of either event precludes the occurrence of the other event.

$$P(A \text{ or } B) = P(A) + P(B)$$

if A and B are mutually exclusive

Example 22. *Three coins are flipped. What is the probability that all three come up heads or all three come up tails?*

Let

$$E = \text{event that all three are heads}$$
$$F = \text{event that all three are tails}$$

Then

$$P(E \text{ or } F) = P(E) + P(F)$$

$$= \frac{1}{2} \cdot \frac{1}{2} \cdot \frac{1}{2} + \frac{1}{2} \cdot \frac{1}{2} \cdot \frac{1}{2} = \frac{1}{8} + \frac{1}{8} = \frac{1}{4} \checkmark$$

The probability that all three coins come up heads was computed as the probability of getting heads on the first coin *and* heads on the second coin *and* heads on the third coin. Each of these probabilities is $\frac{1}{2}$; the ands indicate a product.

Sometimes it is easier to determine the probability that an event will not occur rather than the probability that an event will occur. These two probabilities are related. If E is an event, then E' represents not-E. In other words, $P(E)$ is the probability that E will occur and $P(E')$ is the probability that E will not occur. Since E either occurs or it does not, and E and E' are mutually exclusive,

$$P(E) + P(E') = 1$$

or, after simple algebraic manipulation,

$$P(E) = 1 - P(E')$$

which is the more useful form of the relationship between $P(E)$ and $P(E')$. Here are two examples:

Experiment: Roll two dice.

    E: The sum of the dice is 7.

    E':   The sum of the dice is not 7; that is, the sum is 2, 3, 4, 5, 6, 8, 9, 10, 11, or 12.

Experiment: Draw 4 cards from a standard deck of 52.

    E: Draw four kings.

    E':   Draw anything but four kings; that is, no king, or one king, or two kings, or three kings.

Example 23. *Five coins are flipped. What is the probability that at least one coin will come up heads?*

Let

E = event at least one coin comes up heads

E' = event no coin comes up heads, that is, all five are tails

Then

$$P(E) = 1 - P(E')$$

$$= 1 - \frac{1}{2} \cdot \frac{1}{2} \cdot \frac{1}{2} \cdot \frac{1}{2} \cdot \frac{1}{2}$$

$$= 1 - \frac{1}{32} = \frac{31}{32} \checkmark$$

Note the ease with which the problem is solved in this "indirect" way. If E' were not used, then you would have to determine and add the separate probabilities for obtaining one head, two heads, three heads, four heads, and five heads.

Example 24. *Light bulbs produced by a manufacturer are defective with probability 0.07. If a batch of 10 is selected, what is the probability that exactly 2 of the bulbs chosen are defective?*

Let E = event that a bulb chosen is defective. Then E' is the event that a bulb chosen is good. And

$$P(E) = 0.07$$

$$P(E') = 1 - P(E) = 1 - 0.07 = 0.93$$

The probability we seek is the product of two 0.07 factors (defective) and eight 0.93 factors (good), taking into consideration the $\binom{10}{2}$ or $\binom{10}{8}$ orders possible. Thus,

$$P(\text{exactly two bulbs defective}) = \binom{10}{2}(0.07)^2(0.93)^8 \quad \checkmark$$

EXERCISE SET C

★1. A card is drawn at random from a deck of 52. What is the probability that it is a heart? What is the probability that it is an ace? What is the probability that it is a face card? What is the probability that it is a black card?

2. Two dice are rolled. What is the probability that the sum of the faces of the dice will be 6? 15? greater than 1?

★3. Each question on a multiple-choice test has five possible selections for answers. What is the probability of getting the first three questions correct strictly by guessing?

4. Four coins are flipped. What is the probability that two will come up heads and two will come up tails?

★5. A man draws in turn 4 cards from a deck of 52 without replacement. What is the probability that the first is a black king, the second a red ace, the third an ace, and the fourth a queen?

6. Two cards are chosen without replacement from a deck of 52. What is the probability that both are spades?

★7. Five coins are flipped. What is the probability that all five will come up heads?

8. In a box there are six blue balls and five orange balls. If they are drawn out one by one without replacement, what is the probability that the first will be blue, the second orange, and so on alternately until no balls are left in the box?

★9. A coin has been flipped five times and has come up tails each time. What is the probability that it will come up tails the next time it is flipped?

★10. What is the probability that 5 cards selected from a deck of 52 without replacement will all be diamonds?

11. A bag contains nine white marbles and seven green marbles. If five

*Answers to starred problems are to be found at the back of the book.

marbles are chosen at random without replacement, what is the probability that exactly three of them are white?

*12. From a bowl containing 10 black balls and 7 white balls, 3 balls are picked at random with replacement. What is the probability that 2 are black and 1 is white?

13. A company manufactures light bulbs, 95 per cent of which are good. The remaining 5 per cent are defective. What is the probability that of 10 bulbs selected, exactly 7 are good?

*14. A student has mastered 90 per cent of the test material; that is, the probability of his answering a specific test question correctly is 0.9. If the student takes a 10-question test, what is the probability that he will answer correctly at least 9 of the questions?

*15. A professor has two cars. One of them will start nine-tenths of the time; the other starts eight-tenths of the time. He always tries the more reliable car first. What is the probability that on a given morning he will be able to get a car to start?

16. A basket contains 15 strawberries and 9 cherries. Another basket contains 12 strawberries and 6 cherries. If a basket is selected at random and then one piece of fruit is picked from it, what is the probability that a strawberry will be picked?

*17. Five coins are flipped. What is the probability that at least one of them will land heads?

18. Three dice are rolled. What is the probability that the sum of the three dice is not 9?

chapter ten

---

about calculus

10.1 INTRODUCTION

The study of functions of real numbers leads naturally to the study of calculus. The theory studied in Chapters 1–9 is used extensively in calculus settings. In this chapter, function notation and the concept of slope are applied to develop intuitively the notion of the *derivative* of a function of a real-valued variable. Keep in mind that the development is indeed intuitive. The theory is left for your study in a calculus course. At that time you will apply additionally your knowledge of inequalities and absolute values in order to define and use the concept of *limit*. Finally, keep in mind that this glance at the derivative is but a small sample of the content and nature of a complete course in calculus.

Also in this chapter, you will be introduced to some of the mathematical terminology used in calculus, where formal definitions, theorems, proofs, and corollaries are an integral part of the course.

10.2 THE SLOPE OF A CURVE

The slope of a straight line can be determined by writing the linear equation in $y = mx + b$ form and noting the value of m, the slope.

For any two points on a line, slope can be computed as $\Delta y/\Delta x$, the difference in y coordinate values divided by the difference in x coordinate values. The slope of a line is constant, so it does not matter what two points are used to compute $m = \Delta y/\Delta x$.

But how can the slope of a curve such as $y = x^2$ be determined? Actually, the curve does not have *a* slope. The slope of $y = x^2$ is constantly changing. The slope of the curve at any point is measured by the slope of the tangent line drawn to the curve at the point. The slope of the tangent to $y = x^2$ is 0 at (0, 0).

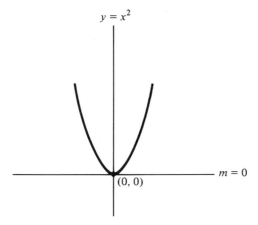

The slope of the tangent is -2 at $(-1, 1)$.

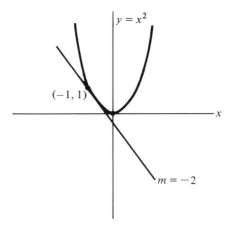

The slope is 4 at (2, 4).

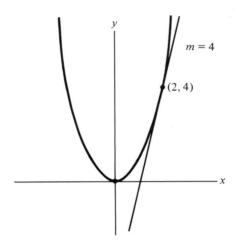

It will be shown that the slope of the (tangent to the) curve $y = x^2$ is itself a function of x. In other words, the value of the slope depends on the x coordinate. Consider $y = x^2$ as $f(x) = x^2$, and consider two points on the curve $y = f(x)$, namely, $(x, f(x))$ and $(x + \Delta x, f(x + \Delta x))$.

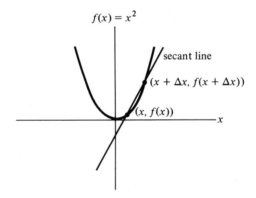

The slope of the line through these two points, called a *secant line*, is

$$m = \frac{\Delta y}{\Delta x} = \frac{\Delta f(x)}{\Delta x} = \frac{f(x + \Delta x) - f(x)}{\Delta x}$$

With $f(x) = x^2$ this becomes

$$m = \frac{(x + \Delta x)^2 - x^2}{\Delta x}$$

$$= \frac{x^2 + 2x(\Delta x) + (\Delta x)^2 - x^2}{\Delta x}$$

$$= \frac{2x(\Delta x) + (\Delta x)^2}{\Delta x}$$

$$= \frac{(2x + \Delta x)\Delta x}{\Delta x}$$

$$= 2x + \Delta x$$

So the slope of the *secant* line through $(x, f(x))$ and $(x + \Delta x, f(x + \Delta x))$ is $m = 2x + \Delta x$; it depends not only on x, but also on Δx. If Δx is made small enough, then the slope $2x + \Delta x$ is very close to $2x$. Furthermore, the secant line through two points is then very nearly the tangent line through $(x, f(x))$.

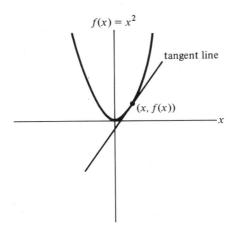

The conclusion suggested is that the slope of the tangent to the curve $f(x) = x^2$ is $2x$.

$$m_{\text{tan}} = 2x$$

For example, the slope of the tangent to $y = x^2$ at $(\underline{2}, 4)$ is $2x$ or $2(\underline{2})$ or 4. Similarly, the slope of the tangent to $y = x^2$ at $(\underline{3}, 9)$ is $2x$ or $2(\underline{3})$ or 6.

Of course, Δx cannot be zero; otherwise, division by zero has been performed. But Δx can be made as close as needed to zero, close enough so that it can be considered zero mechanically in our last step above. A "theory of limits" has been developed to determine precisely the meaning of Δx being "close enough" to zero. Such theory is better left for a calculus course. The notation and mechanics are as follows:

$$\lim_{\Delta x \to 0} (2x + \Delta x) = 2x$$

The limit of $(2x + \Delta x)$ as Δx approaches zero is $2x$.

We have used

$$m_{\text{tan}} = \lim_{\Delta x \to 0} \frac{f(x + \Delta x) - f(x)}{\Delta x}$$

The limit of the difference quotient on the right side of the equals sign above is called the *derivative* of the function $f(x)$ with respect to the variable x. It can be denoted as $f'(x)$. Computing the values of derivatives is a common task in elementary calculus. The process of determining the derivative of a function is called *differentiation*. It is one of two basic calculus operations. The other operation is called *integration*.

10.3 MAXIMUM AND MINIMUM

The curve below has a *relative maximum* functional value at $x = a$ and a *relative minimum* at b. We say "relative" because there are larger and smaller values of the function, but not in the vicinity of a or b.

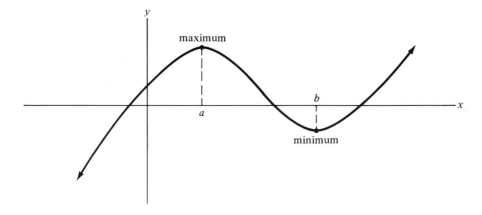

The slope of the tangent to the curve is zero at both the relative maximum and minimum points.

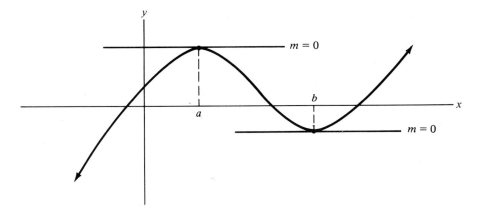

So to find the relative maximum and/or minimum values of such a function, determine where the slope of the tangent is zero. In other words, compute the derivative and set it equal to zero. Solve that equation for the variable.

Example 1. *Find the relative minimum value of* $f(x) = x^2 - 5x$.

Determine $f'(x)$, the derivative of $f(x)$, and set it equal to zero.

$$f'(x) = \lim_{\Delta x \to 0} \frac{f(x + \Delta x) - f(x)}{\Delta x}$$

$$= \lim_{\Delta x \to 0} \frac{[(x + \Delta x)^2 - 5(x + \Delta x)] - [x^2 - 5x]}{\Delta x}$$

$$= \lim_{\Delta x \to 0} \frac{x^2 + 2x(\Delta x) + (\Delta x)^2 - 5x - 5(\Delta x) - x^2 + 5x}{\Delta x}$$

$$= \lim_{\Delta x \to 0} \frac{2x(\Delta x) + (\Delta x)^2 - 5(\Delta x)}{\Delta x}$$

$$= \lim_{\Delta x \to 0} \frac{(2x + \Delta x - 5)\,\Delta x}{\Delta x}$$

$$= \lim_{\Delta x \to 0} (2x + \Delta x - 5)$$

$$= 2x - 5$$

So $f'(x) = 2x - 5$. Setting this equal to zero yields $2x - 5 = 0$ or $x = \frac{5}{2}$. The minimum occurs at $x = \frac{5}{2}$. The value of the function at $\frac{5}{2}$, that is, the relative minimum value of the function, is

$$f\left(\frac{5}{2}\right) = \left(\frac{5}{2}\right)^2 - 5\left(\frac{5}{2}\right) = -\frac{25}{4} \quad \checkmark$$

You might wonder how we know that this is a minimum value rather than a maximum value. Several tests are studied in calculus. One such test is to evaluate the function at points on both sides of the extreme (max or min) point. If x values on both sides yield larger functional values, then the point is a minimum. If both functional values are smaller, then the point is a maximum. If one functional value is larger and one is smaller, then the point is neither a maximum nor a minimum. If the function has more than one extreme point (for example, one maximum and one minimum), points used for the test should be chosen very close to the extreme point. Why?

Example 2. *Find the two numbers whose sum is 20 and whose product is maximum.*

Let the numbers be x and $20 - x$. Represent the product as a function of x; then determine what value of x maximizes that function. The product of the two numbers is $x(20 - x)$ or $20x - x^2$. So determine what value of x maximizes $f(x) = 20x - x^2$.

$$
\begin{aligned}
f'(x) &= \lim_{\Delta x \to 0} \frac{f(x + \Delta x) - f(x)}{\Delta x} \\
&= \lim_{\Delta x \to 0} \frac{[20(x + \Delta x) - (x + \Delta x)^2] - [20x - x^2]}{\Delta x} \\
&= \lim_{\Delta x \to 0} \frac{20x + 20(\Delta x) - x^2 - 2x(\Delta x) - (\Delta x)^2 - 20x + x^2}{\Delta x} \\
&= \lim_{\Delta x \to 0} \frac{20(\Delta x) - 2x(\Delta x) - (\Delta x)^2}{\Delta x} \\
&= \lim_{\Delta x \to 0} \frac{(20 - 2x - \Delta x)\,\Delta x}{\Delta x} \\
&= \lim_{\Delta x \to 0} (20 - 2x - \Delta x) \\
&= 20 - 2x
\end{aligned}
$$

Thus, $f'(x) = 20 - 2x$. The value of x that makes $f'(x) = 0$ is the value that makes $20 - 2x = 0$. In other words, 10 is the x value that maximizes the product function $f(x) = x(20 - x)$. So $x = 10$ and $20 - x = 10$. The maximum product is $10 \cdot 10 = 100$. Note that there is no minimum value for the product. Explain.

EXERCISE SET A

1. Find the derivative of each function.

   ★(a) $f(x) = x$ (b) $f(x) = 5x$

   ★Answers to starred problems are to be found at the back of the book.

⋆(c) $f(x) = 2x + 3$ (d) $f(x) = 3x - 7$

⋆(e) $f(x) = ax$, a constant   ⋆(f) $f(x) = x^2 + 5$

(g) $f(x) = x^2 + 3x$ ⋆(h) $f(x) = 7x^2$

(i) $f(x) = x^2 + 5x + 4$ ⋆(j) $f(x) = x^3$

(k) $f(x) = \sqrt{x}$ (l) $f(x) = \dfrac{1}{x}$

(m) $f(x) = 7$

2. Find the relative maximum and/or minimum values of each function.

⋆(a) $f(x) = x^2 + 2$ (b) $f(x) = x^2 - 6x$

⋆(c) $f(x) = x^2 + 3x + 2$ (d) $f(x) = 3x^2 + 12$

⋆(e) $f(x) = 2x^2 - 9$ (f) $f(x) = 4x^2 + 3x$

(g) $f(x) = x^2 - 3x - 10$ ⋆(h) $f(x) = x^3 - x$

⋆(i) $f(x) = x^3 - x^2$ (j) $f(x) = x^3 - 3x - 4$

3. Find the two numbers whose sum is 50 and whose product is maximum. Then determine the two numbers whose sum is n and whose product is maximum.

⋆4. In problem 6, Exercise Set I, Chapter 2, you had to express the volume of a box as a function of x. Determine the value of x that maximizes the volume of the box. Also, explain what value of x minimizes the volume.

5. Find the width of the window of greatest area, as described in problem 11, Exercise Set I, Chapter 2.

10.4 THEORY IN CALCULUS

Basic to the development of calculus (or any other area of mathematics) are undefined terms, axioms (or postulates), and definitions. An *undefined term* is a word that is assumed to be understood without being defined. Point and line are two words that are usually considered undefined terms. *Axioms* (or *postulates*) are statements that are assumed to be true. They are accepted without proof. The commutative property, $a + b = b + a$ for all real numbers a and b, is usually an axiom in mathematics study. *Definitions* are introduced to extend the foundation built by undefined terms and axioms. Definitions employ undefined terms, axioms, and earlier definitions. For example, the slope of a line can be defined in terms of coordinates of points on the line.

A definition always has two implications. For example, consider the definition of rational number. A *rational number* is a number that can be written in the form a/b, where a and b are integers and b is not zero. This definition tells us two things:

1. If we *have* a rational number "in-hand," then we know just what sort of thing we do have.

2. If we *are looking for* a rational number, then we will know it when we find it.

More specifically, the two implications of this definition are:

1. *If x is rational, then* $x = a/b$, where a and b are integers and $b \neq 0$.

2. If $x = a/b$, where a and b are integers and $b \neq 0$, *then x is* rational.

The implications (if . . . , then . . .) 1 and 2 can be combined by using the expression "if and only if" (often abbreviated *iff*). The result is: x is rational if and only if $x = a/b$, where a and b are integers and $b \neq 0$. Note the two parts suggested by using if and only if.

1. (*If*) x is rational *if* $x = a/b$, where a and b are integers and $b \neq 0$.†

2. (*Only if*) x is rational *only if* $x = a/b$, where a and b are integers and $b \neq 0$.

The "only if" can be rewritten using "if–then," as

2. (*Only if*) If x is rational, then $x = a/b$, where a and b are integers and $b \neq 0$.

In general,

> p if and only if q

means the same as

1. If q, then p.

and

2. If p, then q.

Theorems are statements involving undefined terms, axioms,

---

†The word *if* in the middle of a statement is sometimes replaced by the word *when* or *whenever*.

definitions, and previous theorems. They can be proved by using un-defined terms, axioms, definitions, and earlier theorems. Here is an example of the statement of a theorem.

Theorem: *The equation of the circle with center at* (h, k) *and radius* r *is* $(x - h)^2 + (y - k)^2 = r^2$.

This theorem was proved earlier in the text. There is another theorem that follows immediately from this theorem. Such a "by-product" theorem is often called a *corollary* of the theorem.

Corollary: *The equation of the circle whose center is at the origin and whose radius is* r *is* $x^2 + y^2 = r^2$.

This corollary follows directly from the theorem. Just let $(h, k) = (0, 0)$.

EXERCISE SET B

1. Rewrite each definition in "if and only if" form. Then separate that form into (1) "if" and (2) "only if" statements.
   ⋆(a) A *relation R* is a set of ordered pairs of numbers.
   (b) A *circle* is a set of points whose undirected distances from a fixed point are equal.
   ⋆(c) A *rational function* is a function that can be expressed as the quotient of two polynomial functions.
   (d) The *limit of the function f at a* is L if for every positive number ϵ there is a positive number δ such that $|f(x) - L| < \epsilon$ whenever $0 < |x - a| < \delta$.
   (e) A *function f is continuous at a* if the limit of $f(x)$ as x approaches a is equal to $f(a)$.

2. Assume that each theorem is true and use it to prove the corollary.
   (a) *Theorem:* The graph of $x^2 + y^2 + Dx + Ey + F = 0$ is a circle.
       Corollary: The graph of $Ax^2 + Ay^2 + Dx + Ey + F = 0$ is a circle.
   (b) *Theorem:* If $|x| < a$, then $-a < x < a$.
       Corollary: If $|x - b| < a$, then $b - a < x < b + a$.
   (c) *Theorems:* $D_x 1 = 0$ and $D_x\left(\dfrac{u}{v}\right) = \dfrac{v D_x u - u D_x v}{v^2}$.

   Corollary: $D_x\left(\dfrac{1}{v}\right) = -\dfrac{1}{v^2} D_x v$.

⋆Answers to starred problems are to be found at the back of the book.

appendices

TABLE I   Common Logarithms (base 10)

n	0	1	2	3	4	5	6	7	8	9
1.0	.0000	.0043	.0086	.0128	.0170	.0212	.0253	.0294	.0334	.0374
1.1	.0414	.0453	.0492	.0531	.0569	.0607	.0645	.0682	.0719	.0755
1.2	.0792	.0828	.0864	.0899	.0934	.0969	.1004	.1038	.1072	.1106
1.3	.1139	.1173	.1206	.1239	.1271	.1303	.1335	.1367	.1399	.1430
1.4	.1461	.1492	.1523	.1553	.1584	.1614	.1644	.1673	.1703	.1732
1.5	.1761	.1790	.1818	.1847	.1875	.1903	.1931	.1959	.1987	.2014
1.6	.2041	.2068	.2095	.2122	.2148	.2175	.2201	.2227	.2253	.2279
1.7	.2304	.2330	.2355	.2380	.2405	.2430	.2455	.2480	.2504	.2529
1.8	.2553	.2577	.2601	.2625	.2648	.2672	.2695	.2718	.2742	.2765
1.9	.2788	.2810	.2833	.2856	.2878	.2900	.2923	.2945	.2967	.2989
2.0	.3010	.3032	.3054	.3075	.3096	.3118	.3139	.3160	.3181	.3201
2.1	.3222	.3243	.3263	.3284	.3304	.3324	.3345	.3365	.3385	.3404
2.2	.3424	.3444	.3464	.3483	.3502	.3522	.3541	.3560	.3579	.3598
2.3	.3617	.3636	.3655	.3674	.3692	.3711	.3729	.3747	.3766	.3784
2.4	.3802	.3820	.3838	.3856	.3874	.3892	.3909	.3927	.3945	.3962
2.5	.3979	.3997	.4014	.4031	.4048	.4065	.4082	.4099	.4116	.4133
2.6	.4150	.4166	.4183	.4200	.4216	.4232	.4249	.4265	.4281	.4298
2.7	.4314	.4330	.4346	.4362	.4378	.4393	.4409	.4425	.4440	.4456
2.8	.4472	.4487	.4502	.4518	.4533	.4548	.4564	.4579	.4594	.4609
2.9	.4624	.4639	.4654	.4669	.4683	.4698	.4713	.4728	.4742	.4757
3.0	.4771	.4786	.4800	.4814	.4829	.4843	.4857	.4871	.4886	.4900
3.1	.4914	.4928	.4942	.4955	.4969	.4983	.4997	.5011	.5024	.5038
3.2	.5051	.5065	.5079	.5092	.5105	.5119	.5132	.5145	.5159	.5172
3.3	.5185	.5198	.5211	.5224	.5237	.5250	.5263	.5276	.5289	.5302
3.4	.5315	.5328	.5340	.5353	.5366	.5378	.5391	.5403	.5416	.5428
3.5	.5441	.5453	.5465	.5478	.5490	.5502	.5514	.5527	.5539	.5551
3.6	.5563	.5575	.5587	.5599	.5611	.5623	.5635	.5647	.5658	.5670
3.7	.5682	.5694	.5705	.5717	.5729	.5740	.5752	.5763	.5775	.5786
3.8	.5798	.5809	.5821	.5832	.5843	.5855	.5866	.5877	.5888	.5899
3.9	.5911	.5922	.5933	.5944	.5955	.5966	.5977	.5988	.5999	.6010
4.0	.6021	.6031	.6042	.6053	.6064	.6075	.6085	.6096	.6107	.6117
4.1	.6128	.6138	.6149	.6160	.6170	.6180	.6191	.6201	.6212	.6222
4.2	.6232	.6243	.6253	.6263	.6274	.6284	.6294	.6304	.6314	.6325
4.3	.6335	.6345	.6355	.6365	.6375	.6385	.6395	.6405	.6415	.6425
4.4	.6435	.6444	.6454	.6464	.6474	.6484	.6493	.6503	.6513	.6522
4.5	.6532	.6542	.6551	.6561	.6571	.6580	.6590	.6599	.6609	.6618
4.6	.6628	.6637	.6646	.6656	.6665	.6675	.6684	.6693	.6702	.6712
4.7	.6721	.6730	.6739	.6749	.6758	.6767	.6776	.6785	.6794	.6803
4.8	.6812	.6821	.6830	.6839	.6848	.6857	.6866	.6875	.6884	.6893
4.9	.6902	.6911	.6920	.6928	.6937	.6946	.6955	.6964	.6972	.6981
5.0	.6990	.6998	.7007	.7016	.7024	.7033	.7042	.7050	.7059	.7067
5.1	.7076	.7084	.7093	.7101	.7110	.7118	.7126	.7135	.7143	.7152
5.2	.7160	.7168	.7177	.7185	.7193	.7202	.7210	.7218	.7226	.7235
5.3	.7243	.7251	.7259	.7267	.7275	.7284	.7292	.7300	.7308	.7316
5.4	.7324	.7332	.7340	.7348	.7356	.7364	.7372	.7380	.7388	.7396

TABLE I   (continued)

n	0	1	2	3	4	5	6	7	8	9
5.5	.7404	.7412	.7419	.7427	.7435	.7443	.7451	.7459	.7466	.7474
5.6	.7482	.7490	.7497	.7505	.7513	.7520	.7528	.7536	.7543	.7551
5.7	.7559	.7566	.7574	.7582	.7589	.7597	.7604	.7612	.7619	.7627
5.8	.7634	.7642	.7649	.7657	.7664	.7672	.7679	.7686	.7694	.7701
5.9	.7709	.7716	.7723	.7731	.7738	.7745	.7752	.7760	.7767	.7774
6.0	.7782	.7789	.7796	.7803	.7810	.7818	.7825	.7832	.7839	.7846
6.1	.7853	.7860	.7868	.7875	.7882	.7889	.7896	.7903	.7910	.7917
6.2	.7924	.7931	.7938	.7945	.7952	.7959	.7966	.7973	.7980	.7987
6.3	.7993	.8000	.8007	.8014	.8021	.8028	.8035	.8041	.8048	.8055
6.4	.8062	.8069	.8075	.8082	.8089	.8096	.8102	.8109	.8116	.8122
6.5	.8129	.8136	.8142	.8149	.8156	.8162	.8169	.8176	.8182	.8189
6.6	.8195	.8202	.8209	.8215	.8222	.8228	.8235	.8241	.8248	.8254
6.7	.8261	.8267	.8274	.8280	.8287	.8293	.8299	.8306	.8312	.8319
6.8	.8325	.8331	.8338	.8344	.8351	.8357	.8363	.8370	.8376	.8382
6.9	.8388	.8395	.8401	.8407	.8414	.8420	.8426	.8432	.8439	.8445
7.0	.8451	.8457	.8463	.8470	.8476	.8482	.8488	.8494	.8500	.8506
7.1	.8513	.8519	.8525	.8531	.8537	.8543	.8549	.8555	.8561	.8567
7.2	.8573	.8579	.8585	.8591	.8597	.8603	.8609	.8615	.8621	.8627
7.3	.8633	.8639	.8645	.8651	.8657	.8663	.8669	.8675	.8681	.8686
7.4	.8692	.8698	.8704	.8710	.8716	.8722	.8727	.8733	.8739	.8745
7.5	.8751	.8756	.8762	.8768	.8774	.8779	.8785	.8791	.8797	.8802
7.6	.8808	.8814	.8820	.8825	.8831	.8837	.8842	.8848	.8854	.8859
7.7	.8865	.8871	.8876	.8882	.8887	.8893	.8899	.8904	.8910	.8915
7.8	.8921	.8927	.8932	.8938	.8943	.8949	.8954	.8960	.8965	.8971
7.9	.8976	.8982	.8987	.8993	.8998	.9004	.9009	.9015	.9020	.9025
8.0	.9031	.9036	.9042	.9047	.9053	.9058	.9063	.9069	.9074	.9079
8.1	.9085	.9090	.9096	.9101	.9106	.9112	.9117	.9122	.9128	.9133
8.2	.9138	.9143	.9149	.9154	.9159	.9165	.9170	.9175	.9180	.9186
8.3	.9191	.9196	.9201	.9206	.9212	.9217	.9222	.9227	.9232	.9238
8.4	.9243	.9248	.9253	.9258	.9263	.9269	.9274	.9279	.9284	.9289
8.5	.9294	.9299	.9304	.9309	.9315	.9320	.9325	.9330	.9335	.9340
8.6	.9345	.9350	.9355	.9360	.9365	.9370	.9375	.9380	.9385	.9390
8.7	.9395	.9400	.9405	.9410	.9415	.9420	.9425	.9430	.9435	.9440
8.8	.9445	.9450	.9455	.9460	.9465	.9469	.9474	.9479	.9484	.9489
8.9	.9494	.9499	.9504	.9509	.9513	.9518	.9523	.9528	.9533	.9538
9.0	.9542	.9547	.9552	.9557	.9562	.9566	.9571	.9576	.9581	.9586
9.1	.9590	.9595	.9600	.9605	.9609	.9614	.9619	.9624	.9628	.9633
9.2	.9638	.9643	.9647	.9652	.9657	.9661	.9666	.9671	.9675	.9680
9.3	.9685	.9689	.9694	.9699	.9703	.9708	.9713	.9717	.9722	.9727
9.4	.9731	.9736	.9741	.9745	.9750	.9754	.9759	.9763	.9768	.9773
9.5	.9777	.9782	.9786	.9791	.9795	.9800	.9805	.9809	.9814	.9818
9.6	.9823	.9827	.9832	.9836	.9841	.9845	.9850	.9854	.9859	.9863
9.7	.9868	.9872	.9877	.9881	.9886	.9890	.9894	.9899	.9903	.9908
9.8	.9912	.9917	.9921	.9926	.9930	.9934	.9939	.9943	.9948	.9952
9.9	.9956	.9961	.9965	.9969	.9974	.9978	.9983	.9987	.9991	.9996

TABLE II   Trigonometric Functions of Angles in *Radians*

t	$\sin t$	$\cos t$	$\tan t$	$\cot t$	$\sec t$	$\csc t$
.00	.0000	1.0000	.0000	. . . . .	1.000	. . . . .
.01	.0100	1.0000	.0100	99.997	1.000	100.00
.02	.0200	.9998	.0200	49.993	1.000	50.00
.03	.0300	.9996	.0300	33.323	1.000	33.34
.04	.0400	.9992	.0400	24.987	1.001	25.01
.05	.0500	.9988	.0500	19.983	1.001	20.01
.06	.0600	.9982	.0601	16.647	1.002	16.68
.07	.0699	.9976	.0701	14.262	1.002	14.30
.08	.0799	.9968	.0802	12.473	1.003	12.51
.09	.0899	.9960	.0902	11.081	1.004	11.13
.10	.0998	.9950	.1003	9.967	1.005	10.02
.11	.1098	.9940	.1104	9.054	1.006	9.109
.12	.1197	.9928	.1206	8.293	1.007	8.353
.13	.1296	.9916	.1307	7.649	1.009	7.714
.14	.1395	.9902	.1409	7.096	1.010	7.166
.15	.1494	.9888	.1511	6.617	1.011	6.692
.16	.1593	.9872	.1614	6.197	1.013	6.277
.17	.1692	.9856	.1717	5.826	1.015	5.911
.18	.1790	.9838	.1820	5.495	1.016	5.586
.19	.1889	.9820	.1923	5.200	1.018	5.295
.20	.1987	.9801	.2027	4.933	1.020	5.033
.21	.2085	.9780	.2131	4.692	1.022	4.797
.22	.2182	.9759	.2236	4.472	1.025	4.582
.23	.2280	.9737	.2341	4.271	1.027	4.386
.24	.2377	.9713	.2447	4.086	1.030	4.207
.25	.2474	.9689	.2553	3.916	1.032	4.042
.26	.2571	.9664	.2660	3.759	1.035	3.890
.27	.2667	.9638	.2768	3.613	1.038	3.749
.28	.2764	.9611	.2876	3.478	1.041	3.619
.29	.2860	.9582	.2984	3.351	1.044	3.497
.30	.2955	.9553	.3093	3.233	1.047	3.384
.31	.3051	.9523	.3203	3.122	1.050	3.278
.32	.3146	.9492	.3314	3.018	1.053	3.179
.33	.3240	.9460	.3425	2.920	1.057	3.086
.34	.3335	.9428	.3537	2.827	1.061	2.999
.35	.3429	.9394	.3650	2.740	1.065	2.916
.36	.3523	.9359	.3764	2.657	1.068	2.839
.37	.3616	.9323	.3879	2.578	1.073	2.765
.38	.3709	.9287	.3994	2.504	1.077	2.696
.39	.3802	.9249	.4111	2.433	1.081	2.630
.40	.3894	.9211	.4228	2.365	1.086	2.568
.41	.3986	.9171	.4346	2.301	1.090	2.509
.42	.4078	.9131	.4466	2.239	1.095	2.452
.43	.4169	.9090	.4586	2.180	1.100	2.399
.44	.4259	.9048	.4708	2.124	1.105	2.348

TABLE II (continued)

t	sin t	cos t	tan t	cot t	sec t	csc t
.45	.4350	.9004	.4831	2.070	1.111	2.299
.46	.4439	.8961	.4954	2.018	1.116	2.253
.47	.4529	.8916	.5080	1.969	1.122	2.208
.48	.4618	.8870	.5206	1.921	1.127	2.166
.49	.4706	.8823	.5334	1.875	1.133	2.125
.50	.4794	.8776	.5463	1.830	1.139	2.086
.51	.4882	.8727	.5594	1.788	1.146	2.048
.52	.4969	.8678	.5726	1.747	1.152	2.013
.53	.5055	.8628	.5859	1.707	1.159	1.978
.54	.5141	.8577	.5994	1.668	1.166	1.945
.55	.5227	.8525	.6131	1.631	1.173	1.913
.56	.5312	.8473	.6269	1.595	1.180	1.883
.57	.5396	.8419	.6310	1.560	1.188	1.853
.58	.5480	.8365	.6552	1.526	1.196	1.825
.59	.5564	.8309	.6696	1.494	1.203	1.797
.60	.5646	.8253	.6841	1.462	1.212	1.771
.61	.5729	.8196	.6989	1.431	1.220	1.746
.62	.5810	.8139	.7139	1.401	1.229	1.721
.63	.5891	.8080	.7291	1.372	1.238	1.697
.64	.5972	.8021	.7445	1.343	1.247	1.674
.65	.6052	.7961	.7602	1.315	1.256	1.652
.66	.6131	.7900	.7761	1.288	1.266	1.631
.67	.6210	.7838	.7923	1.262	1.276	1.610
.68	.6288	.7776	.8087	1.237	1.286	1.590
.69	.6365	.7712	.8253	1.212	1.297	1.571
.70	.6442	.7648	.8423	1.187	1.307	1.552
.71	.6518	.7584	.8595	1.163	1.319	1.534
.72	.6594	.7518	.8771	1.140	1.330	1.517
.73	.6669	.7452	.8949	1.117	1.342	1.500
.74	.6743	.7358	.9131	1.095	1.354	1.483
.75	.6816	.7317	.9316	1.073	1.367	1.467
.76	.6889	.7248	.9505	1.052	1.380	1.452
.77	.6961	.7179	.9697	1.031	1.393	1.437
.78	.7033	.7109	.9893	1.011	1.407	1.422
.79	.7104	.7038	1.009	.9908	1.421	1.408
.80	.7174	.6967	1.030	.9712	1.435	1.394
.81	.7243	.6895	1.050	.9520	1.450	1.381
.82	.7311	.6822	1.072	.9331	1.466	1.368
.83	.7379	.6749	1.093	.9146	1.482	1.355
.84	.7446	.6675	1.116	.8964	1.498	1.343
.85	.7513	.6600	1.138	.8785	1.515	1.331
.86	.7578	.6524	1.162	.8609	1.533	1.320
.87	.7643	.6448	1.185	.8437	1.551	1.308
.88	.7707	.6372	1.210	.8267	1.569	1.297
.89	.7771	.6294	1.235	.8100	1.589	1.287

TABLE II (continued)

t	$\sin t$	$\cos t$	$\tan t$	$\cot t$	$\sec t$	$\csc t$
.90	.7833	.6216	1.260	.7936	1.609	1.277
.91	.7895	.6137	1.286	.7774	1.629	1.267
.92	.7956	.6058	1.313	.7615	1.651	1.257
.93	.8016	.5978	1.341	.7458	1.673	1.247
.94	.8076	.5898	1.369	.7303	1.696	1.238
.95	.8134	.5817	1.398	.7151	1.719	1.229
.96	.8192	.5735	1.428	.7001	1.744	1.221
.97	.8249	.5653	1.459	.6853	1.769	1.212
.98	.8305	.5570	1.491	.6707	1.795	1.204
.99	.8360	.5487	1.524	.6563	1.823	1.196
1.00	.8415	.5403	1.557	.6421	1.851	1.188
1.01	.8468	.5319	1.592	.6281	1.880	1.181
1.02	.8521	.5234	1.628	.6142	1.911	1.174
1.03	.8573	.5148	1.665	.6005	1.942	1.166
1.04	.8624	.5062	1.704	.5870	1.975	1.160
1.05	.8674	.4976	1.743	.5736	2.010	1.153
1.06	.8724	.4889	1.784	.5604	2.046	1.146
1.07	.8772	.4801	1.827	.5473	2.083	1.140
1.08	.8820	.4713	1.871	.5344	2.122	1.134
1.09	.8866	.4625	1.917	.5216	2.162	1.128
1.10	.8912	.4536	1.965	.5090	2.205	1.122
1.11	.8957	.4447	2.014	.4964	2.249	1.116
1.12	.9001	.4357	2.066	.4840	2.295	1.111
1.13	.9044	.4267	2.120	.4718	2.344	1.106
1.14	.9086	.4176	2.176	.4596	2.395	1.101
1.15	.9128	.4085	2.234	.4475	2.448	1.096
1.16	.9168	.3993	2.296	.4356	2.504	1.091
1.17	.9208	.3902	2.360	.4237	2.563	1.086
1.18	.9246	.3809	2.427	.4120	2.625	1.082
1.19	.9284	.3717	2.498	.4003	2.691	1.077
1.20	.9320	.3624	2.572	.3888	2.760	1.073
1.21	.9356	.3530	2.650	.3773	2.833	1.069
1.22	.9391	.3436	2.733	.3659	2.910	1.065
1.23	.9425	.3342	2.820	.3546	2.992	1.061
1.24	.9458	.3248	2.912	.3434	3.079	1.057
1.25	.9490	.3153	3.010	.3323	3.171	1.054
1.26	.9521	.3058	3.113	.3212	3.270	1.050
1.27	.9551	.2963	3.224	.3102	3.375	1.047
1.28	.9580	.2867	3.341	.2993	3.488	1.044
1.29	.9608	.2771	3.467	.2884	3.609	1.041
1.30	.9636	.2675	3.602	.2776	3.738	1.038
1.31	.9662	.2579	3.747	.2669	3.878	1.035
1.32	.9687	.2482	3.903	.2562	4.029	1.032
1.33	.9711	.2385	4.072	.2456	4.193	1.030
1.34	.9735	.2288	4.256	.2350	4.372	1.027

TABLE II **(continued)**

t	$\sin t$	$\cos t$	$\tan t$	$\cot t$	$\sec t$	$\csc t$
1.35	.9757	.2190	4.455	.2245	4.566	1.025
1.36	.9779	.2092	4.673	.2140	4.779	1.023
1.37	.9799	.1994	4.913	.2035	5.014	1.021
1.38	.9819	.1896	5.177	.1931	5.273	1.018
1.39	.9837	.1798	5.471	.1828	5.561	1.017
1.40	.9854	.1700	5.798	.1725	5.883	1.015
1.41	.9871	.1601	6.165	.1622	6.246	1.013
1.42	.9887	.1502	6.581	.1519	6.657	1.011
1.43	.9901	.1403	7.055	.1417	7.126	1.010
1.44	.9915	.1304	7.602	.1315	7.667	1.009
1.45	.9927	.1205	8.238	.1214	8.299	1.007
1.46	.9939	.1106	8.989	.1113	9.044	1.006
1.47	.9949	.1006	9.887	.1011	9.938	1.005
1.48	.9959	.0907	10.938	.0910	11.029	1.004
1.49	.9967	.0807	12.350	.0810	12.390	1.003
1.50	.9975	.0707	14.101	.0709	14.137	1.003
1.51	.9982	.0608	16.428	.0609	16.458	1.002
1.52	.9987	.0508	19.670	.0508	19.965	1.001
1.53	.9992	.0408	24.498	.0408	24.519	1.001
1.54	.9995	.0308	32.461	.0308	32.476	1.000
1.55	.9998	.0208	48.078	.0208	48.089	1.000
1.56	.9999	.0108	92.620	.0108	92.626	1.000
1.57	1.0000	.0008	1255.8	.0008	1255.8	1.000
1.58	1.0000	−.0092	−108.65	−.0092	−108.65	1.000
1.59	.9998	−.0192	−52.067	−.0192	−52.08	1.000
1.60	.9996	−.0292	−34.233	−.0292	−34.25	1.000

TABLE III Trigonometric Functions of Angles in *Degrees*

Angle	Sin	Cos	Tan
0°	.0000	1.0000	.0000
1°	.0175	.9998	.0175
2°	.0349	.9994	.0349
3°	.0523	.9986	.0524
4°	.0698	.9976	.0699
5°	.0872	.9962	.0875
6°	.1045	.9945	.1051
7°	.1219	.9925	.1228
8°	.1392	.9903	.1405
9°	.1564	.9877	.1584
10°	.1736	.9848	.1763
11°	.1908	.9816	.1944
12°	.2079	.9781	.2126
13°	.2250	.9744	.2309
14°	.2419	.9703	.2493
15°	.2588	.9659	.2679
16°	.2756	.9613	.2867
17°	.2924	.9563	.3057
18°	.3090	.9511	.3249
19°	.3256	.9455	.3443
20°	.3420	.9397	.3640
21°	.3584	.9336	.3839
22°	.3746	.9272	.4040
23°	.3907	.9205	.4245
24°	.4067	.9135	.4452
25°	.4226	.9063	.4663
26°	.4384	.8988	.4877
27°	.4540	.8910	.5095
28°	.4695	.8829	.5317
29°	.4848	.8746	.5543
30°	.5000	.8660	.5774
31°	.5150	.8572	.6009
32°	.5299	.8480	.6249
33°	.5446	.8387	.6494
34°	.5592	.8290	.6745
35°	.5736	.8192	.7002
36°	.5878	.8090	.7265
37°	.6018	.7986	.7536
38°	.6157	.7880	.7813
39°	.6293	.7771	.8098
40°	.6428	.7660	.8391
41°	.6561	.7547	.8693
42°	.6691	.7431	.9004
43°	.6820	.7314	.9325
44°	.6947	.7193	.9657
45°	.7071	.7071	1.0000

TABLE III (continued)

Angle	Sin	Cos	Tan
46°	.7193	.6947	1.0355
47°	.7314	.6820	1.0724
48°	.7431	.6691	1.1106
49°	.7547	.6561	1.1504
50°	.7660	.6428	1.1918
51°	.7771	.6293	1.2349
52°	.7880	.6157	1.2799
53°	.7986	.6018	1.3270
54°	.8090	.5878	1.3764
55°	.8192	.5736	1.4281
56°	.8290	.5592	1.4826
57°	.8387	.5446	1.5399
58°	.8480	.5299	1.6003
59°	.8572	.5150	1.6643
60°	.8660	.5000	1.7321
61°	.8746	.4848	1.8040
62°	.8829	.4695	1.8807
63°	.8910	.4540	1.9626
64°	.8988	.4384	2.0503
65°	.9063	.4226	2.1445
66°	.9135	.4067	2.2460
67°	.9205	.3907	2.3559
68°	.9272	.3746	2.4751
69°	.9336	.3584	2.6051
70°	.9397	.3420	2.7475
71°	.9455	.3256	2.9042
72°	.9511	.3090	3.0777
73°	.9563	.2924	3.2709
74°	.9613	.2756	3.4874
75°	.9659	.2588	3.7321
76°	.9703	.2419	4.0108
77°	.9744	.2250	4.3315
78°	.9781	.2079	4.7046
79°	.9816	.1908	5.1446
80°	.9848	.1736	5.6713
81°	.9877	.1564	6.3138
82°	.9903	.1392	7.1154
83°	.9925	.1219	8.1443
84°	.9945	.1045	9.5144
85°	.9962	.0872	11.4301
86°	.9976	.0698	14.3007
87°	.9986	.0523	19.0811
88°	.9994	.0349	28.6363
89°	.9998	.0175	57.2900
90°	1.0000	.0000	

Answers to Starred Exercises

1. (a) Rational, integer, whole, natural   (c) Rational

   (e) Irrational   (g) Rational

   (i) Rational, integer, whole, natural   (l) Rational

4. (a) $(2 + a)x$ (b) $y^2(y - 7)$

   (d) $(x + 10)(x - 10)$ (f) $(2 + m)(x + y)$

   (h) $(a - b)^2$ (i) $(x + 6)(x + 1)$

   (l) $2(b + 2)(b - 2)$ (o) $a^3(a + 3)(a - 3)$

   (q) $(x + 1)(x - 1)(x^2 + 1)$ (s) $(m - 3)(m^2 + 3m + 9)$

5. (a) $\dfrac{bx + ay}{ab}$ (c) $\dfrac{15 + 7x^2}{3xy}$

   (d) $\dfrac{x - 26}{(x + 2)(x - 5)}$ (f) $\dfrac{7x^2 - 14x + 15}{(x^2 - 1)(x^2 - 7x + 10)}$

   (g) $\dfrac{2x}{x + 1}$ (i) $\dfrac{x - 4}{x^2}$

   (k) $\dfrac{3x(10y + 1)}{2(x + 1)}$ (m) $\dfrac{2}{x - 2}$

   (o) $\dfrac{2x + 1}{3x - 1}$ (p) $\dfrac{x^2 + y^2}{x^2}$

(q) $\dfrac{x+2y}{x+4y}$ 　　　　　　　　　　(s) $\dfrac{t+1}{1-t}$

(u) $\dfrac{-1}{x^2-2x+hx-h+1}$

6. (a) $x=\dfrac{16}{7}$ 　　　　　　　　(c) $x=-2$

(e) $x=21$ 　　　　　　　　　(f) $x=-\dfrac{1}{4}$

(i) $x=\dfrac{11}{2}$ 　　　　　　　　(k) $x=-\dfrac{5}{3}$

11. (a) $\dfrac{73}{99}$ 　(b) $\dfrac{5}{9}$ 　(d) $\dfrac{24}{45}$

13. (a) Commutative 　(b) Distributive 　(c) Associative
(d) Commutative
(e) Distributive. (Also, $5\cdot2=10$ can be considered closure, too.)

Page 18

Chapter 1　EXERCISE SET B

1. (a) $12m^{17}$ 　　　　(b) $4x^{12}y^7$ 　　　　(c) $x^{10}y^{15}$

(f) 33 　　　　　　(i) $\dfrac{x^8y^6}{2}$ 　　　　(j) $\dfrac{7n^6r^5s^2}{p^2q^2}$

(k) $\dfrac{y^4}{x^8}$ 　　　　(n) $\dfrac{a^3+b^3}{ab}$ 　　　(o) $\dfrac{c^5+d^3}{c^3d^2}$

(p) 0 　　　　　(s) $\dfrac{a+b}{ab}$ 　　　　(v) $\dfrac{xy}{x+y}$

(w) $\dfrac{b^{2y}-a^{2x}}{a^xb^y}$ 　　(y) $\dfrac{x^2y^2}{x+y}$ 　　　(z) $\dfrac{1}{1-x^4}$

2. (a) $\dfrac{x-1}{x^3}$ 　　　　　　　　(c) $\dfrac{x}{x+1}$

Page 22-24

Chapter 1　EXERCISE SET C

1. (a) 12 　　　　　(b) 1000 　　　　(c) 5

(d) 9 　　　　　(h) 10 　　　　(i) $\dfrac{1}{4}$

(l) $2\sqrt{5}$ 　　　　(o) $2x^3y^2\sqrt{3}$ 　　(p) $2\sqrt[3]{3}$

(r) $4x^6$ 　$8x^7$ 　(s) $-2n$ 　　　　(w) $\dfrac{3}{2}$

(z) $\dfrac{\sqrt{2}}{2}$

2. (a) $\dfrac{3\sqrt{7}}{7}$ (b) $\dfrac{5\sqrt{2}}{6}$ (d) $-3(1+\sqrt{2})$

(f) $5(\sqrt{3}+\sqrt{2})$ (h) $\dfrac{5\sqrt[3]{49}}{7}$ (k) $\sqrt{7}$

3. (a) $\dfrac{3x+2}{\sqrt{x+1}}$ (c) $\dfrac{x^2(9-11x^2)}{(1-x^2)^{2/3}}$

(e) $\dfrac{2(11x^2-15)}{(x^2-3)^{2/5}}$ (f) $\dfrac{2x^3(x^2-2)}{(x^2-1)^{3/2}}$

4. (a) $x=\dfrac{63}{2}$ (b) No solution

(d) $x=7$ (f) $x=42$

(g) $x=621$ (j) $x=\dfrac{19}{10}$

(k) $x=\dfrac{30}{17}$

7. (a) Zero when $x=\frac{1}{3}$; undefined for $x=\pm3$.
 (b) Zero when $x=0$.
 (c) Undefined for $x=\pm2\sqrt{2}$.
 (e) Zero when $x=0,\ -\frac{2}{3}$; undefined for $x=3$.
 (f) Zero when $x=-1$; undefined for $x=0$.

10. (a) $\dfrac{2\sqrt[3]{49}}{7}$ (c) $\dfrac{5\sqrt[3]{2}}{4}$

Chapter 1   EXERCISE SET D   *Page 28-29*

1. (a) $x=-4,-1$ (b) $x=3,4$
  (c) $y=5,-2$ (e) $x=2,-7$

  (f) $t=3,-3$ (g) $m=-2,-\dfrac{3}{2}$

  (h) $x=\dfrac{2}{3},-5$

2. (a) $x=-1,-5$ (b) $x=-4\pm3\sqrt{2}$
  (c) $n=2\pm\sqrt{3}$ (d) $m=1\pm\sqrt{2}$

  (f) $y=\dfrac{-1\pm\sqrt{5}}{2}$ (i) $x=\dfrac{-3\pm\sqrt{41}}{4}$

3. (a) $x=-1,-3$ (b) $x=2,5$

  (c) $t=1\pm\sqrt{2}$ (e) $x=\dfrac{-3\pm\sqrt{37}}{2}$

Page 29

(g) $x = 1 \pm \sqrt{6}$

(h) $x = \dfrac{-5 \pm 5\sqrt{5}}{4}$

4. $x = 4, -\dfrac{5}{2}$

Page 38-39

Chapter 1   EXERCISE SET E

1. $x < 5$

3. $x < \dfrac{3}{2}$

7. $x \geq 6$

10. $x \geq \dfrac{9}{2}$

12. $x > \dfrac{11}{3}$

14. $x > 6$

17. $x \geq 5$ or $x \leq -2$

19. $x > 1$ or $x < -1$

20. $2 < x < 8$

21. $x > 2$

24. $x < -1$

25. $x > 0$ or $x < -5$

26. $-\sqrt{7} < x < \sqrt{7}$

27. $x < 0$ or $1 < x < 3$

29. $x > 3$ or $-3 < x < 0$

30. $-5 < x < 3$

32. $x > 7$ or $-2 < x < 2$

33. True for all the real numbers.

Page 44

Chapter 1   EXERCISE SET F

1. $-6 < x < 6$

3. $-4 < n < 4$

5. $-\dfrac{13}{7} < x < \dfrac{13}{7}$

7. $-8 < m < 4$

10. $x = -5$

11. $-5 < x < \dfrac{5}{3}$

13. $-\dfrac{8}{7} \leq z \leq \dfrac{16}{7}$

14. $-11 < x < 4$

17. $-\dfrac{\epsilon}{3} < x < \dfrac{\epsilon}{3}$

18. $6 - \epsilon < x < 6 + \epsilon$

19. $x > 7$ or $x < -7$

21. $w > 7$ or $w < -7$

23. $x > 54$ or $x < -36$

25. $x > \dfrac{23}{2}$ or $x < -\dfrac{9}{2}$

28. $x > -\dfrac{7}{2}$ or $x < -\dfrac{17}{2}$

37. $\delta = 0.005$

38. (a) No solution; impossible for $|7x + 2|$ to be negative.

(b) True for all real numbers, since $|5x - 3|$ is always nonnegative.

Chapter 1   EXERCISE SET G   *Page 48-49*

1. (a) $-i$ (c) 1 (e) $-3i$

(g) $3i$ (j) $i\sqrt{3}$ (l) $2i\sqrt{3}$

(n) $9i\sqrt{3}$

2. (a) $x = \dfrac{-1 \pm i\sqrt{3}}{2}$ (d) $x = 1 \pm 2i$

(e) $x = \dfrac{-3 \pm 3i\sqrt{3}}{2}$ (g) $x = \dfrac{1 \pm 2i\sqrt{29}}{3}$

(i) $x = \pm 2i\sqrt{2}$

3. (a) $7 + 4i$ (c) $13 - 3i$ (e) $14 - 3i$

(g) $-9 + 19i$ (i) $63 + 16i$ (l) $2 + 11i$

(m) $1 - i$ (o) $-\dfrac{1}{2} + \dfrac{1}{2}i$ (q) $\dfrac{1}{5} + \dfrac{3}{5}i$

Chapter 2   EXERCISE SET A   *Page 52*

1. (a) True (b) False (c) False

(d) False (e) False (f) False

2. (a) $\{x \mid x \text{ real}, x > 13\}$ (c) $\{x \mid x \text{ real}, x \geq 0\}$

(d) $\{x \mid x \text{ real}, -5 \leq x \leq 16\}$ (g) $\{x \mid x \text{ even}, 4 \leq x \leq 300\}$

(h) $\{x \mid x \text{ real}, -50 < x < 400, x \neq 0\}$

3. (a) $[2, 5]$ (b) $(-5, 7]$ (c) $[0, 6)$

(d) $(-4, 0)$ (e) $(0, \infty)$ (f) $(-\infty, -9]$

Chapter 2   EXERCISE SET B   *Page 62*

1. a, b, c, e, f, j are functions.

2. D: all the reals; R: all the reals

3. D: $x \geq 0$; R: $y \geq 0$ **6.** D: $x \neq 0$; R: $y \neq 0$

7. D: $x \neq 9$; R: $y \neq 0$ **8.** D: $\{3, 5, 6\}$; R: $\{5, 3, -3\}$

9. D: $x \geq 3$; R: $y \geq 0$ **12.** D: all the reals; R: $y \geq 1$

14. D: $x \geq 0$; R: all the reals

15. D: $-4 \leq x \leq 4$; R: $-4 \leq y \leq 4$

18. D: $x \geq 0$; R: $y \leq 0$ **20.** D: $x \geq 4$; R: $y \leq 0$

22. D: $x \neq 0, -1$; R: $y > 0, y \leq -16$

23. D: $x \geq 4, x \leq -4$; R: all the reals

28. D: all the reals; R: $-\sqrt{13} < y < 0, 0 < y < \sqrt{13}$

29. D: all the reals; R: $y \geq -\dfrac{9}{4}$

32. D: $-\dfrac{7}{2\sqrt{3}} \leq x \leq \dfrac{7}{2\sqrt{3}}$; R: all the reals

34. D: $x \neq 0$; R: $y \neq 0$

page 65

Chapter 2   EXERCISE SET C

1. (b) 1   (c) -5 (d) -20

(e) $3a - 5$ (f) $-\dfrac{7}{2}$ (h) $9m - 5$

(j) $3x - 20$

2. (a) 19   (d) -1 (e) -8

(g) $t^2 + 8t - 1$ (h) $\dfrac{13}{4}$ (i) $-\dfrac{47}{16}$

(j) $x^2 + 12x + 19$

3. (a) 1   (c) $2x + \Delta x$ (e) 0

(g) $3x^2 + 3x\Delta x + (\Delta x)^2$ (i) $\dfrac{-1}{x(x + \Delta x)}$

5. Even: b, d, g; odd: a, f; neither: c, e, h, i, j

7. (a) $f + g = x + 3, f - g = x - 3, f \cdot g = 3x, \dfrac{f}{g} = \dfrac{x}{3}$

Chapter 2   EXERCISE SET D

1. D: all the reals; R: $f(x) \geq 0$ (no asymptotes)

2. D: all the reals; R: 5 only   (no asymptotes)

3. D: all the reals; R: all the reals   (no asymptotes)

6. D: all the reals; R: $f(x) \geq 0$ (no asymptotes)

8. D: $x \geq 0$; R: $f(x) \geq 0$ (no asymptotes)

9. D: all the reals; R: $f(x) \leq 2$ (no asymptotes)

11. D: all the reals; R: $h(x) \leq 0$ (no asymptotes)

14. D: all the reals; R: all the reals   (no asymptotes)

15. D: $x \neq -2$; R: $f(x) \neq 3$ (no asymptotes)

18. D: all the reals; R: nonnegative integers   (no asymptotes)

20. D: all the reals; R: $F(x) \leq 0$ (no asymptotes)

Chapter 2   EXERCISE SET E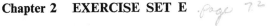

1. $f(g(x)) = x^2 + 6x + 9; g(f(x)) = x^2 + 3$

3. $f(g(x)) = \dfrac{1}{x^2}; g(f(x)) = \dfrac{1}{x^2}$

5. $f(g(x)) = \dfrac{2 - x}{3 - x}; g(f(x)) = \dfrac{x + 2}{x + 1}$

6. $f(g(x)) = |2x + 5|; g(f(x)) = 2|x| + 5$

8. $f(g(x)) = \dfrac{3x}{2x + 1}; g(f(x)) = \dfrac{x + 2}{3}$

10. $f(g(x)) = \dfrac{1 + 2x}{1 - 2x}; g(f(x)) = \dfrac{x - 2}{x + 2}$

Chapter 2   EXERCISE SET F

1. $\{(3, 0), (4, 1), (8, 2), (6, -3)\}$ (function)

2. $\{(2, 1), (4, 1), (8, 1)\}$ (function)

3. $\{(2, 3), (-8, 4), (-2, 6), (2, 10)\}$ (not a function)

4. $y = \dfrac{x + 3}{5}$ (function)

8. $y = \pm\sqrt{x + 3}$ (not a function)

9. $y = \sqrt[3]{x}$ (function)

13. $y = \dfrac{2x+3}{x}$ (function)

14. $y = x^2 - 5$ (function)

16. $y = \dfrac{3}{x^2+1}$ (function)

17. $y = \dfrac{-3 \pm \sqrt{4x-11}}{2}$ (not a function)

Chapter 2   EXERCISE SET G

1. Symmetric with respect to the y axis.

2. Symmetric with respect to the x axis.

3. Symmetric with respect to the origin.

5. Symmetric with respect to the y axis.

8. Symmetric with respect to the origin.

9. Symmetric with respect to the x axis, y axis, and the origin.

Chapter 2   EXERCISE SET I

1. $V(d) = \dfrac{\pi d^3}{6}$ **2.** $V(h) = \dfrac{4\pi h^3}{3}$

3. $P(x) = 2x + \dfrac{50}{x}$ **6.** $V(x) = (10 - 2x)^2 x$

7. $A(P) = \dfrac{P^2}{4\pi}$ **9.** $a = \dfrac{b^2}{10}$

11. $A = \dfrac{32r - \pi r^2 - 4r^2}{2}$

Chapter 3   EXERCISE SET A

2. (a) $x = 4$ (c) $x = -3$ (e) $x = 1$

(f) $x = 1$ (h) $x = -6$ (j) $x = -\dfrac{1}{4}$

(n) $x = -\dfrac{1}{2}$ (o) $x = -2$ (r) $x = \dfrac{5}{2}$

(t) $x = -1$

Page 95

Chapter 3   EXERCISE SET B

1. (a) $\log_3 81 = 4$ (b) $\log_5 25 = 2$ (e) $\log_9 3 = \dfrac{1}{2}$

(g) $\log_4 \dfrac{1}{4} = -1$ (i) $\log_2 b = a$

2. (a) $10^2 = 100$ (c) $25^{1/2} = 5$ (e) $3^{-2} = \dfrac{1}{9}$

3. (a) 4   (c) 10   (e) $\dfrac{1}{3}$ (f) $\dfrac{1}{5}$

(h) -2 (i) 1   (k) 2   (l) x

Chapter 3   EXERCISE SET C

Page 97-98

1. (a) 12   (d) 10   (f) 25   (h) 2

(i) -3 (j) -2 (k) 3.5

2. (a) $\log_t 35$ (c) $\log_b 750$ (d) $\log_a 3.5$ (e) $\log_b 576$

3. (a) $\log_b 7x$ (b) $\log_b x^3$

Chapter 3   EXERCISE SET D

Page 102-104

1. (a) 1.5866   (b) 2.6911   (c) 0.7324

(f) $9.9212 - 10$ (i) $7.8129 - 10$

2. (a) 258   (b) 85.8   (c) 1420

(d) 0.552   (g) 0.00650   (h) 3.04

3. (a) 2.6895   (b) 3.5334   (c) 1.4717

(e) 3.3647   (i) $6.1735 - 10$

4. (a) 75.33   (b) 0.5535   (e) 276.1

(f) 0.03378   (i) 4.362

8. (a) $x = 10, x = 100$

10. $x = \ln 2$

Chapter 3   EXERCISE SET E

Page 106

1. (a) 1.732   (c) 4.918   (e) 398.2

(g) 6.155   (i) 697.7   (j) 231.0

(k) 4.704   (l) 4190   (m) 294.1

(o) 57.70

3. Approximately 17,700,000 miles.

Page 108

Chapter 3   EXERCISE SET F

1. (a) $x \doteq 1.465$ (c) $x \doteq -0.192$ (e) $x \doteq 1.709$

(g) $x \doteq -0.464$

2. (a) $x = 2$ (b) $x = 70$

3. (a) $x < 0.6309$ (approximately)

4. (a) $x = \log \dfrac{8}{3}$ (b) $x = \dfrac{\log (c/a)}{\log b}$ or $x = \log_b (c/a)$

Page 109

Chapter 3   EXERCISE SET G

1. (a) 2.66   (c) 2.27   (e) 3.25

(g) 3.06   (i) 1.20

2. $\log_3 549 = \dfrac{\log_8 549}{\log_8 3}$

3. $\log_b N = \dfrac{\log_B N}{\log_B b}$

Chapter 4   EXERCISE SET A

1. (a) $(-1, 0)$ (b) $(0, -1)$ (d) $\left(-\dfrac{1}{2}, \dfrac{\sqrt{3}}{2}\right)$

(f) $(1, 0)$ (i) $\left(\dfrac{1}{2}, \dfrac{\sqrt{3}}{2}\right)$ (l) $\left(-\dfrac{\sqrt{2}}{2}, \dfrac{\sqrt{2}}{2}\right)$

(m) $\left(-\dfrac{\sqrt{2}}{2}, -\dfrac{\sqrt{2}}{2}\right)$

2. $\dfrac{x}{2}\sqrt{3}$

3. $x\sqrt{2}$

Chapter 4   EXERCISE SET B

1.

t	0	$\dfrac{\pi}{6}$	$\dfrac{\pi}{4}$	$\dfrac{\pi}{3}$	$\dfrac{\pi}{2}$	π	$\dfrac{3\pi}{2}$
$\cos t$	1	$\dfrac{\sqrt{3}}{2}$	$\dfrac{\sqrt{2}}{2}$	$\dfrac{1}{2}$	0	-1	0
$\sin t$	0	$\dfrac{1}{2}$	$\dfrac{\sqrt{2}}{2}$	$\dfrac{\sqrt{3}}{2}$	1	0	-1
$\tan t$	0	$\dfrac{1}{\sqrt{3}}$	1	$\sqrt{3}$	$-*$	0	$-$
$\cot t$	$-$	$\sqrt{3}$	1	$\dfrac{1}{\sqrt{3}}$	0	$-$	0
$\sec t$	1	$\dfrac{2}{\sqrt{3}}$	$\sqrt{2}$	2	$-$	-1	$-$
$\csc t$	$-$	2	$\sqrt{2}$	$\dfrac{2}{\sqrt{3}}$	1	$-$	-1

* Dash indicates undefined.

Chapter 4   EXERCISE SET C

21. $\sin t = \dfrac{-\sqrt{21}}{5}$, $\tan t = \dfrac{\sqrt{21}}{2}$, $\cot t = \dfrac{2}{\sqrt{21}}$, $\csc t = \dfrac{-5}{\sqrt{21}}$, $\sec t = -\dfrac{5}{2}$

Chapter 4   EXERCISE SET D

1. (a) 0.8290   (b) 0.9781   (c) 19.0811   (d) 0.6745
  (e) 0.5000   (f) 0.7660   (g) 0.9135   (h) 0.8290
  (i) 0.4848   (j) 0.7660   (k) 0.5150   (l) 0.9397

2. (a) 0.9320   (b) 0.8776   (c) 0.3429   (d) 0.0108
  (e) 0.2341   (f) 0.3429   (g) 0.4350   (h) -0.4259
  (i) 0.2579   (j) -0.5480

3. (a) 0.9536   (b) 0.7812   (c) 0.6402   (d) 0.4323
  (e) 0.9977   (f) 0.5327   (g) 0.6440   (h) 0.7011
  (i) 0.7888   (j) 0.3417

4. (a) $-\dfrac{1}{2}$ (b) $\dfrac{\sqrt{2}}{2}$ (c) $-\dfrac{\sqrt{3}}{2}$ (d) $\dfrac{1}{2}$

(e) $-\dfrac{1}{2}$ (f) $-\dfrac{\sqrt{3}}{2}$ (g) $-\dfrac{\sqrt{2}}{2}$ (h) $\dfrac{1}{2}$

(i) $-\dfrac{1}{2}$ (j) $\dfrac{\sqrt{3}}{2}$ (k) $-\dfrac{1}{2}$ (l) $-\dfrac{1}{2}$

(m) $-\dfrac{\sqrt{3}}{2}$ (n) $\dfrac{\sqrt{2}}{2}$ (o) $-\dfrac{1}{2}$ (p) $\dfrac{\sqrt{2}}{2}$

(q) $\dfrac{\sqrt{2}}{2}$ (r) $-\dfrac{\sqrt{3}}{2}$ (s) $-\dfrac{1}{2}$ (t) $-\dfrac{\sqrt{3}}{2}$

5. (a) $\dfrac{3\pi}{2}$ (b) $\dfrac{5\pi}{12}$ (c) $36°$ (d) $30°$

(e) $18°$ (f) $\dfrac{3\pi}{4}$ (g) $\dfrac{2\pi}{3}$ (h) $\dfrac{7\pi}{6}$

(i) $135°$ (j) $\dfrac{7\pi}{4}$ (k) 5π (l) $\dfrac{5\pi}{4}$

(m) $540°$ (n) $210°$ (o) $585°$

Chapter 4   EXERCISE SET E

2. (a) Period $= 2\pi$; amplitude $= 1$
(b) Period $= 2\pi$; amplitude $= 1$
(c) Period $= \pi$; amplitude not defined
(d) Period $= \pi$; amplitude not defined
(e) Period $= 2\pi$; amplitude not defined
(f) Period $= 2\pi$; amplitude not defined
(g) Period $= 2\pi$; amplitude $= 2$
(h) Period $= 2\pi$; amplitude $= \dfrac{1}{2}$
(i) Period $= 2\pi$; amplitude $= 6$
(j) Period $= \dfrac{\pi}{2}$; amplitude $= 1$
(k) Period $= \pi$; amplitude $= 1$
(l) Period $= \dfrac{2\pi}{3}$; amplitude $= 1$
(m) Period $= \pi$; amplitude $= 1$
(n) Period $= 2\pi$; amplitude $= 5$
(o) Period $= \pi$; amplitude $= 3$

(p) Period $= 4\pi$; amplitude $= 2$

(q) Period $= \dfrac{2\pi}{3}$; amplitude $= 4$

(r) Period $= 8\pi$; amplitude $= 2$

(s) Period $= 2\pi$; amplitude $= 1$

(t) Period $= 2\pi$; amplitude $= 1$

(u) Period $= 2\pi$; amplitude $= 1$

(v) Period $= 2\pi$; amplitude $= 1$

(w) Period $= 2\pi$; amplitude $= 1$

(x) Period $= 2\pi$; amplitude $= 1$

Chapter 4   EXERCISE SET F

1. $\dfrac{\sqrt{6} - \sqrt{2}}{4}$

4. (a) $\sin 3t = 3 \sin t - 4 \sin^3 t$

   (b) $\cos 3t = 4 \cos^3 t - 3 \cos t$

5. (a) $\dfrac{63}{65}$ (b) $\dfrac{16}{65}$ (e) $\dfrac{63}{16}$

   (f) $\dfrac{24}{25}$ (h) $-\dfrac{7}{25}$

6. $\cot (u + v) = \dfrac{\cot u \cot v - 1}{\cot u + \cot v}$, $\cot 2t = \dfrac{\cot^2 t - 1}{2 \cot t}$

7. $\sin \dfrac{\theta}{2} = +\dfrac{1}{\sqrt{10}}$, $\cos \dfrac{\theta}{2} = -\dfrac{3}{\sqrt{10}}$, $\tan \dfrac{\theta}{2} = -\dfrac{1}{3}$

Chapter 4   EXERCISE SET G

1. $\dfrac{\pi}{6}$ 2. $\dfrac{\pi}{3}$ 3. $\dfrac{5\pi}{6}$

7. $\dfrac{\sqrt{2}}{2}$ 8. $\dfrac{\sqrt{3}}{2}$ 10. $\dfrac{1}{2}$

13. 0.9987 14. 0.9664 15. $\dfrac{2}{\sqrt{5}}$

17. $\dfrac{1}{\sqrt{5}}$ 19. $\dfrac{3}{4}$

Chapter 4   EXERCISE SET H

1 $x = \dfrac{\pi}{2} + 2n\pi$ (n = any integer)

4. $x = \dfrac{\pi}{3} + n\pi$; $x = \dfrac{2\pi}{3} + n\pi$ (n = any integer)

5. $x = \dfrac{3\pi}{2} + 2n\pi$ (n = any integer)

7. $x = \dfrac{\pi}{3} + n\pi$; $x = \dfrac{2\pi}{3} + n\pi$ (n = any integer)

10. $u = \pm\dfrac{\pi}{4} + n\pi$ (n = any integer)

11. No solution

13. $x = 2n\pi$; $x = \dfrac{3\pi}{2} + 2n\pi$ (n = any integer)

16. $x = (2n + 1)\pi$; $x = \text{Cos}^{-1}\left(\dfrac{1}{3}\right) + 2n\pi$;

$x = 2\pi - \text{Cos}^{-1}\left(\dfrac{1}{3}\right) + 2n\pi$ (n = any integer)

Chapter 4   EXERCISE SET I

1. $\sin x = \dfrac{2}{\sqrt{13}}$, $\cos x = \dfrac{3}{\sqrt{13}}$, $\tan x = \dfrac{2}{3}$,

$\sin y = \dfrac{3}{\sqrt{13}}$, $\cos y = \dfrac{2}{\sqrt{13}}$, $\tan y = \dfrac{3}{2}$

3. (a) $x \doteq 1.4$ (b) $x \doteq 2.8$ (d) $x \doteq 7.5$

4. (a) $\theta \doteq 37°$ (b) $\theta \doteq 34°$ (c) $\theta \doteq 55°$ (d) $\theta \doteq 24°$

Chapter 4   EXERCISE SET J

1. (a) $\left(2\sqrt{2}, \dfrac{\pi}{4}\right)$ (b) $(5, 0)$

   (c) $\left(2, \dfrac{\pi}{6}\right)$ (d) $\left(2, -\dfrac{\pi}{3}\right)$

(e) $(-5, 0)$ (f) $(0, 0)$

(g) $\left(-1, -\dfrac{\pi}{4}\right)$ (h) $(\sqrt{6.3716},\ \text{Tan}^{-1}\ 0.77)$

2. (a) $\left(\dfrac{\sqrt{3}}{2}, \dfrac{1}{2}\right)$ (b) $(2\sqrt{2}, 2\sqrt{2})$

(c) $(3, 0)$ (d) $(-3, 0)$

(e) $(0, 2)$ (f) $(0, 0)$

(g) $(-1, 0)$ (h) $(\cos 1,\ \sin 1)$

4. (a) $r = \dfrac{-3}{\sin \theta - 2 \cos \theta}$ (b) $r^2 = 25$

(d) $r^2 = \dfrac{16}{\cos 2\theta}$ (e) $x^2 + y^2 - 6y = 0$

(h) $x^2 + y^2 = 25$ (i) $x = 3$

(j) $r(\sin \theta - r \cos^2 \theta) = 0$ (l) $y = \dfrac{x}{\sqrt{3}}$

(n) $x^2 + y^2 = \left[\text{Tan}^{-1}\left(\dfrac{y}{x}\right)\right]^2$

(p) $x^4 + y^4 - 2y^3 + 2x^2 y^2 - 2x^2 y - x^2 = 0$

Chapter 4   EXERCISE SET K

1. (a) $2(\cos 30° + i \sin 30°)$ (b) $1(\cos 0° + i \sin 0°)$

(c) $1(\cos 90° + i \sin 90°)$ (g) $2(\cos 60° + i \sin 60°)$

(h) $2\sqrt{2}(\cos 45° + i \sin 45°)$ (j) $\sqrt{13}(\cos 34° - i \sin 34°)$

2. (a) $3i$ (c) $\sqrt{2} + i\sqrt{2}$ (e) $\dfrac{\sqrt{2}}{2} - \dfrac{i\sqrt{2}}{2}$

(f) -3 (h) $-\sqrt{2} - i\sqrt{2}$

3. (a) $3(\cos 60° + i \sin 60°)$

(c) $35\left(\cos \dfrac{3\pi}{4} + i \sin \dfrac{3\pi}{4}\right)$

(e) $2\left(\cos \dfrac{17\pi}{12} + i \sin \dfrac{17\pi}{12}\right)$

4. (a) $3(\cos 30° + i \sin 30°)$

(c) $\dfrac{5}{7}\left(\cos \dfrac{\pi}{4} + i \sin \dfrac{\pi}{4}\right)$

(e) $\dfrac{1}{2}\left(\cos \dfrac{\pi}{4} - i \sin \dfrac{\pi}{4}\right)$

5. (a) $81(\cos 120° + i \sin 120°)$
(c) $2401(\cos 180° + i \sin 180°)$
(d) $32(\cos 300° + i \sin 300°)$
(f) $16(\cos 1080° - i \sin 1080°)$

6. (a) $1, -\dfrac{1}{2} \pm \dfrac{i\sqrt{3}}{2}$ (c) $2, -1 \pm i\sqrt{3}$

(d) $i, \pm\dfrac{\sqrt{3}}{2} - \dfrac{i}{2}$

(g) $2^{1/8}(\cos 11.25° + i \sin 11.25°)$, $2^{1/8}(\cos 101.25° + i \sin 101.25°)$,
$2^{1/8}(\cos 191.25° + i \sin 191.25°)$, $2^{1/8}(\cos 281.25° + i \sin 281.25°)$

Chapter 4   EXERCISE SET L

1. $x \doteq 5.7$ **2.** $x \doteq 8.1$ **4.** $x \doteq 30°$ **5.** $x \doteq 10.2$

Chapter 5   EXERCISE SET A

1. $-3, -4$ **3.** $\pm 4i$

4. $\dfrac{-1 \pm i\sqrt{11}}{2}$ **6.** $0, -2, -5$

8. -9 (double zero)   **11.** $\pm 3, -2, -4$ **12.** $\pm 1, \pm i$

14. (a) 48 ft/sec   (b) 2176 ft   (c) 15 sec
      (d) -400 ft/sec   (e) -80 ft/sec   (f) 2500 ft

Chapter 5   EXERCISE SET B

1. (a) Quotient: $x^2 + 5x + 2$;   remainder: 0
(b) Quotient: $x^2 + 4x - 5$;   remainder: 0
(f) Quotient: $x^2 - 4x + 12$;   remainder: -23
(g) Quotient: $2x^3 + 4x^2 + 5x + 3$;   remainder: 9
(j) Quotient: $x^3 - 2x^2 + 4x - 10$;   remainder: 21
(k) Quotient: $x^2 + x + 1$;   remainder: 0
(l) Quotient: $x^2 + xy + y^2$;   remainder: 0

2. (a) $(x - 2)(x - 3)(x + 1)$ (b) $x(x - 4)(x - 5)$
(c) $(x - 1)(x + 3)(x + 2)$ (g) $(x + 4)(2x + 1)(x + 5)$
(i) $(x + 4)(3x + 5)(x - 2)$

3. (a) $5, \pm i$ (c) $-2, 1 \pm i\sqrt{3}$
   (e) $2, -3 \pm i$ (g) $-5, -1 \pm 2i$
   (h) $1, -3, -\dfrac{1}{2}$ (i) $-2, 7, \pm i$
   (l) $3, -\dfrac{1}{2}, -1 \pm i\sqrt{6}$

4. (a) $2, 3, -1$ (b) $3, \pm 2i$ (d) $1, -\dfrac{1}{2}, \dfrac{-1 \pm i\sqrt{3}}{2}$

7. (a) From $(x - 4)(x + 1)(x - 1) > 0$ $x > 4$ or $-1 < x < 1$

Chapter 6   EXERCISE SET A

Page 212–214

1. (a) $4x + 3y + 5 = 0$ (c) $x - 7y - 2 = 0$
   (e) $x + 9y - 4 = 0$ (g) $3x - 1 = 0$
   (i) $6x - y - 4 = 0$

2. (a) $\dfrac{1}{2}$ (c) 3 (e) 5 (g) $-\dfrac{7}{9}$

3. (a) $x - 2y + 9 = 0$ (c) $3x - y - 10 = 0$
   (e) $y = 5x$ (g) $7x + 9y + 13 = 0$

4. $\sqrt{3}$

6. $y = 0$; the slope is zero.

8. $y = 2x - 7$

10. (a) $(2, 7)$ (b) $(1, 3)$ (c) $\left(-\dfrac{1}{2}, -3\right)$ (d) $(-4, 3\tfrac{1}{2})$ (e) $(-2\tfrac{1}{2}, 1\tfrac{1}{2})$

11. (a) $x + 2y - 15 = 0$ (b) $4x + 2y - 19 = 0$
    (c) $x + y - 6 = 0$ (d) $x - y - 1 = 0$
    (e) $3x - y + 13 = 0$ (f) $x = \dfrac{7}{2}$

12. (a) $m = 2, b = -1$ (b) $m = -1, b = 4$
    (c) $m = 1, b = 2$ (e) $m = 3, b = 3$
    (f) $m = -1, b = 0$ (g) $m = 2, b = -6$
    (i) $m = \dfrac{1}{3}, b = \dfrac{4}{3}$ (k) $m = 2, b = -3$
    (o) $m = -\dfrac{2}{3}, b = \dfrac{1}{3}$

15. $x = -3$ **19.** $y = -3x + 6$

Page 217

Chapter 6   EXERCISE SET B

1. (a) 5   (b) 12   (c) $\sqrt{29}$

2. (a) $\dfrac{17}{5}$ (c) $\dfrac{6}{\sqrt{29}}$ (e) $\dfrac{17}{\sqrt{5}}$ (g) $\dfrac{5}{\sqrt{2}}$

page 228 - 229

Chapter 6   EXERCISE SET C

1. $x^2 + y^2 = 1$

2. (a) $(x - 3)^2 + (y - 12)^2 = 16$
(b) $(x - 1)^2 + y^2 = 4$
(c) $(x + 2)^2 + (y + 7)^2 = 100$
(d) $x^2 + (y + 9)^2 = 1$

3. (a) Center: $(4, 5)$;   radius: 7
(b) Center: $(9, -8)$;   radius: 8
(c) Center: $(-1, -2)$;   radius: $\sqrt{5}$
(d) Center: $(0, 1)$;   radius: $2\sqrt{3}$

4. (a) Center: $(-2, -3)$;   radius: 1
(b) Center: $(-6, 5)$;   radius: 6
(c) Center: $(3, 1)$;   radius: 8

6. (a) Vertex: $(0, 0)$;   focus: $(2, 0)$;   directrix: $x = -2$
(d) Vertex: $(0, 0)$;   focus: $(0, -3)$;   directrix: $y = 3$

(e) Vertex: $(0, 0)$;   focus: $\left(\dfrac{5}{2}, 0\right)$;   directrix: $x = -\dfrac{5}{2}$

(g) Vertex: $(0, 0)$;   focus: $\left(0, \dfrac{1}{2}\right)$;   directrix: $y = -\dfrac{1}{2}$

(i) Vertex: $(0, 0)$;   focus: $\left(-\dfrac{1}{6}, 0\right)$;   directrix: $x = \dfrac{1}{6}$

7. (a) $y^2 = 20x$ (c) $x^2 = 2y$ (e) $y^2 = 4x$

8. (a) Vertices: $(0, \pm 1), (\pm 3, 0)$;   foci: $(\pm 2\sqrt{2}, 0)$
(c) Vertices: $(0, \pm\sqrt{10}), (\pm 4, 0)$;   foci: $(\pm\sqrt{6}, 0)$
(e) Vertices: $(\pm 3, 0)$;   foci: $(\pm\sqrt{13}, 0)$

(g) Vertices: $(\pm 2, 0)$, $(0, \pm 3)$;   foci: $(0, \pm\sqrt{5})$

(h) Vertices: $(0, \pm 2)$;   foci: $(0, \pm\sqrt{13})$

Chapter 6   EXERCISE SET D

1. (a) $\dfrac{(x-2)^2}{9} + \dfrac{(y+1)^2}{4} = 1$ (ellipse)

Vertices: $(-1, -1)$, $(5, -1)$, $(2, 1)$, $(2, -3)$; foci: $(2 \pm \sqrt{5}, -1)$

(c) $\dfrac{(x-3)^2}{4} - \dfrac{y^2}{25} = 1$ (hyperbola)

Vertices: $(5, 0)$, $(1, 0)$; foci: $(3 \pm \sqrt{29}, 0)$

(e) $y^2 = 7(x+3)$ (parabola)

Vertex: $(-3, 0)$; directrix: $x = -\dfrac{19}{4}$

(f) $\dfrac{(x+2)^2}{1} + \dfrac{(y-1)^2}{4} = 1$ (ellipse)

Vertices: $(-3, 1)$, $(-1, 1)$ $(-2, 3)$, $(-2, -1)$; foci: $(-2, 1 \pm \sqrt{3})$

(g) $\dfrac{x^2}{16} - \dfrac{(y+5)^2}{9} = 1$ (hyperbola)

Vertices: $(4, -5)$, $(-4, -5)$; foci: $(5\ -5)$, $(-5, -5)$

(h) $(y+3)^2 = 9(x+2)$ (parabola)

Vertex: $(-2, -3)$; directrix: $x = -\dfrac{17}{4}$

2. (a) Ellipse   (b) Hyperbola   (c) Parabola   (d) Hyperbola

(e) Ellipse

Chapter 7   EXERCISE SET A

Page 241

1. (a) $\left(\dfrac{2}{3}, \dfrac{7}{3}\right)$ (b) $(1, 1)$ (c) $\left(\dfrac{1}{2}, 1\right)$

(d) $\left(\dfrac{1}{3}, \dfrac{8}{3}\right)$ (i) $(-3, 2)$

2. (a) $x = 3, y = 5$ (b) $x = 4, y = 1$

(e) $x = 6, y = 2$ (f) $x = 7, y = -2$

(i) $x = -5, y = -4$

3. (a) $x = 5, y = 1$ (c) $x = 4, y = -1$

(d) $x = 7, y = -2$ (f) $x = 4, y = -3$

(g) $x = 3, y = -4$

4. (a) $x = 1, y = 2, z = 3$ (c) $x = 2, y = 0, z = -3$

    (e) $x = 1, y = -2, z = 4$ (g) $r = 2, s = -3, t = 4$

    (i) $x = 2, y = -3, z = -1$

Chapter 7   EXERCISE SET B

1. (a) 10   (b) 0   (c) 12

    (d) 0   (e) -19 (f) 72

    (g) 0   (h) 0   (i) 0

    (j) 1

2. (a) $x = 3, y = 4$ (b) $x = 0, y = 0$

    (c) $x = -2, y = -5$ (e) $p = \dfrac{2}{3}, q = -\dfrac{5}{3}$

    (h) $m = \dfrac{143}{47}, n = -\dfrac{221}{47}$

3. (a) 123   (b) 28   (c) 0   (d) -64 (f) -113

4. (a) $x = \dfrac{4}{5}, y = \dfrac{7}{5}, z = \dfrac{16}{5}$ (b) $x = -15, y = 6, z = 3$

    (c) $x = \dfrac{32}{9}, y = -\dfrac{7}{6}, z = \dfrac{65}{18}$ (e) $x = \dfrac{583}{248}, y = \dfrac{33}{248}, z = \dfrac{167}{248}$

    (f) $r = \dfrac{247}{195}, s = -\dfrac{15}{195}, t = \dfrac{361}{195}$

Chapter 7   EXERCISE SET C

1. (a) 4   (b) 93   (d) 0   (g) 2304   (h) 364

2. (a) $w = 6, x = -1, y = 0, z = 4$

    (c) $A = 1, B = 0, D = -1, F = 1$

    (e) $r = 6, s = 2, t = 3, u = -3$

Chapter 7   EXERCISE SET D

1. (a) $x = 1, y = 3$ (d) $x = 2, y = 4$

    (f) $u = -2, v = 3$ (g) $x = 1, y = -\dfrac{1}{2}$

    (i) $x = 1, y = 1, z = 3$ (k) $x = 5, y = 1, z = 2$

    (n) $x = 0, y = -5, z = 2$ (o) $t = -7, u = -6, v = 2$

Chapter 7   EXERCISE SET E

1. (a) $x^2 + y^2 - 6x + 4y - 12 = 0$
   (b) $x^2 + y^2 - 4x + 10y - 20 = 0$

2. (a) $\dfrac{2}{x} + \dfrac{7}{x+5}$ (d) $\dfrac{9}{x+5} - \dfrac{3}{x-3} + \dfrac{1}{x-7}$

   (e) $\dfrac{4}{x+2} + \dfrac{7x-1}{x^2+5x+1}$ (f) $\dfrac{\frac{3}{5}}{x-3} + \dfrac{6}{x-6} - \dfrac{\frac{3}{5}}{x+2}$

Chapter 7   EXERCISE SET F   page 273

1. (a) $(3, 4), (-4, -3)$ (b) No solution
   (d) $(0, 2)$ (f) $(4, 0), (-4, 0)$

   (h) $\left(\pm\dfrac{2\sqrt{15}}{3}, \dfrac{\sqrt{21}}{3}\right), \left(\pm\dfrac{2\sqrt{15}}{3}, -\dfrac{\sqrt{21}}{3}\right)$

   (i) $\left(\dfrac{1+\sqrt{73}}{4}, \dfrac{9-3\sqrt{73}}{8}\right), \left(\dfrac{1-\sqrt{73}}{4}, \dfrac{9+3\sqrt{73}}{8}\right)$

2. $2 + i, -2 - i$

3. $\left(\dfrac{5+\sqrt{13}}{2}, \dfrac{-1+\sqrt{13}}{2}\right), \left(\dfrac{5-\sqrt{13}}{2}, \dfrac{-1-\sqrt{13}}{2}\right)$

4. (a) $(2, 3), (-1, 0)$ (c) $(\sqrt{5}, 1), (-\sqrt{5}, 1)$
   (e) $(0, 0), (1, 0), (5, 0)$

Chapter 8   EXERCISE SET A

1. (a) $a_n = n + 1$ (b) $a_n = 3n$
   (c) $a_n = \dfrac{1}{n}$ (d) $a_n = -2n$
   (g) $a_n = (n + 2)^2$ (i) $a_n = \dfrac{5}{n^2}$

2. (a) $4, 7, 10, 13$ (b) $-3, -1, 1, 3$
   (d) $-1, 8, 23, 44$ (g) $0, \dfrac{4}{3}, 4, 8$
   (h) $\dfrac{1}{2}, \dfrac{4}{3}, \dfrac{9}{4}, \dfrac{16}{5}$

3. (a) 47 (b) 575 (c) 217
   (e) 7200 (f) -24 (h) $98, 1648$

4. (a) 640   (b) 2916   (c) $\dfrac{3}{256}$

   (d) 10^8 (e) $\dfrac{1093}{729}$ (h) 264

   (j) 3   (k) 160   (m) $4(3)^{n-1}$ or
                                        $-4(-3)^{n-1}$

5. (a) 8   (b) $\dfrac{3}{2}$ (f) $\dfrac{3}{8}$

   (h) $\dfrac{343}{6}$ (i) No finite sum

7. (a) 2.71665   (b) 2.71804   (c) 2.71824

8. 0.6823

10. (a) 2.89524   (b) 3.33968   (c) 2.97605   (d) 3.28374

Chapter 8   EXERCISE SET C

1. (a) 1.414   (e) 3.316   (g) 10.344

2. (a) 2.15   (c) 3.80   (e) 4.64

Chapter 9   EXERCISE SET A

Page 296

1. 5040   **2.** 90   **5.** 64·625

6. 676,000   **8.** $\dfrac{11!}{2!2!2!}$ **9.** $\dfrac{20!}{7!9!4!}$

11. (a) 15   (b) 45   (c) 32   (d) 165   (e) 1   (f) 1

13. $\dbinom{14}{5}$

15. $\dbinom{23}{6}\dbinom{17}{17}$ or just $\dbinom{23}{6}$

17. $18 \cdot \dbinom{17}{3}$

18. $\dbinom{9}{1}\dbinom{6}{2} + \dbinom{9}{2}\dbinom{6}{1} + \dbinom{9}{1}\dbinom{6}{3} + \dbinom{9}{2}\dbinom{6}{2} + \dbinom{9}{3}\dbinom{6}{1}$

page 303   *16*

Chapter 9   EXERCISE SET B

1. (a) 285   (b) 120   (e) 127   (g) $\frac{71}{20}$

2. (a) $c^5 + 5c^4d + 10c^3d^2 + 10c^2d^3 + 5cd^4 + d^5$

(d) $a^3 - 3a^2b + 3ab^2 - b^3$

(f) $32x^5 + 80x^4y + 80x^3y^2 + 40x^2y^3 + 10xy^4 + y^5$

(h) $32x^5 - 240x^4y + 720x^3y^2 - 1080x^2y^3 + 810xy^4 - 243y^5$

(k) 1.061520150601

(m) $1 - 7x^2 + 21x^4 - 35x^6 + 35x^8 - 21x^{10} + 7x^{12} - x^{14}$

3. $1710x^4y^{18}$

Chapter 9   EXERCISE SET C

1. $\frac{1}{4}, \frac{1}{13}, \frac{3}{13}, \frac{1}{2}$

3. $\frac{1}{125}$

5. $\frac{2}{52} \cdot \frac{2}{51} \cdot \frac{3}{50} \cdot \frac{4}{49}$

7. $\left(\frac{1}{2}\right)^5$

9. $\frac{1}{2}$

10. $\frac{13}{52} \cdot \frac{12}{51} \cdot \frac{11}{50} \cdot \frac{10}{49} \cdot \frac{9}{48}$

12. $\binom{3}{2}\left(\frac{10}{17}\right)^2\left(\frac{7}{17}\right)$

14. $\binom{10}{9}(0.9)^9(0.1) + (0.9)^{10}$

15. $\frac{9}{10} + \frac{1}{10} \cdot \frac{8}{10}$

17. $1 - \left(\frac{1}{2}\right)^5$

Chapter 10   EXERCISE SET A

1. (a) 1   (c) 2   (e) a (f) $2x$ (h) $14x$ (j) $3x^2$

2. (a) $f(0) = 2$ is minimum

(c) $f\left(-\frac{3}{2}\right) = -\frac{1}{4}$ is minimum

(e) $f(0) = -9$ is minimum

(h) $f\left(\frac{\sqrt{3}}{3}\right) = -\frac{2\sqrt{3}}{9}$ is minimum

$f\left(-\frac{\sqrt{3}}{3}\right) = \frac{2\sqrt{3}}{9}$ is maximum

(i) $f(0) = 0$ is maximum

$f\left(\dfrac{2}{3}\right) = -\dfrac{4}{27}$ is minimum

4. The value 5/3 maximizes the volume. The value 5 minimizes the volume, since it creates a "box" with volume equal to zero.

Chapter 10   EXERCISE SET B

1. (a) R is a relation if and only if it is a set of ordered pairs of numbers.

      (1) R is a relation *if* it is a set of ordered pairs of numbers.

      (2) R is a relation *only if* it is a set of ordered pairs of numbers.

  or (2) If R is a relation, then it is a set of ordered pairs of numbers.

  (c) A function is a rational function if and only if it can be expressed as the quotient of two polynomial functions.

      (1) A function is rational *if* it can be expressed as the quotient of two polynomial functions.

      (2) A function is rational *only if* it can be expressed as the quotient of two polynomial functions.

  or (2) If a function is rational, then it can be expressed as the quotient of two polynomial functions.

Index